Medicinal Plants of India

Medicinal Plants of India

Preeti Dhar
SUNY New Paltz, USA

Durga Nath Dhar
Indian Institute of Technology, India

World Scientific

NEW JERSEY · LONDON · SINGAPORE · BEIJING · SHANGHAI · HONG KONG · TAIPEI · CHENNAI · TOKYO

Published by

World Scientific Publishing Co. Pte. Ltd.

5 Toh Tuck Link, Singapore 596224

USA office: 27 Warren Street, Suite 401-402, Hackensack, NJ 07601

UK office: 57 Shelton Street, Covent Garden, London WC2H 9HE

British Library Cataloguing-in-Publication Data

A catalogue record for this book is available from the British Library.

MEDICINAL PLANTS OF INDIA

ISBN 978-981-120-338-1

For any available supplementary material, please visit
https://www.worldscientific.com/worldscibooks/10.1142/11358#t=suppl

Typeset by Stallion Press
Email: enquiries@stallionpress.com

Disclaimer

The authors and the publisher, World Scientific Publishing, shall have neither liability nor responsibility to any person or entity with respect to any loss, damage or injury caused or alleged to be caused directly or indirectly by the information in this book. The information presented herein is not intended to substitute or replace medical counseling.

Preface

Our knowledge of medicinal plants is the outcome of careful observations made over thousands of years. There has been a recent surge in the interest of medicinal plants among researchers worldwide. The basic aim of these researchers is to identify and synthesize the bioactive molecules found in the medicinal plants. In the United States, a screening program funded by the National Cancer Institute led to the discovery of the bioactive molecule — taxol, obtained from the bark of *Taxus brevifolia*. Taxol was found to be very effective against several cancers, but patients required large amounts of it for the treatment and it was available only in small amounts from natural sources. Later, researchers were able to obtain sizeable amounts of taxol by a semi-synthetic method from the leaves of *Taxus baccata*. There are many other examples in the chemical literature wherein medicinal plants have served as a source of bioactive molecules that are being used by practitioners of modern medicine.

There are multiple advantages to using naturally occurring bioactive molecules. First, the human body is endowed with a wonderful mechanism of identifying a naturally occurring bioactive molecule in comparison to a synthetic heterocyclic compound that is not found in nature. Such synthetic compounds have harmful side effects in our system. Second, some naturally occurring compounds derived from medicinal plants are adaptogenic. This property results in a balanced impact of the compound in the human body. An example is the antidiabetic herb, gurmar. In sharp contrast to synthetic oral antidiabetic drugs, gurmar lowers only higher than normal blood sugar levels and does not lower blood sugar to dangerously low levels. Third, most medicinal plants are very affordable and can also be easily cultivated.

This book provides an overview of the common uses of many popular medicinal plants that are found worldwide and outlines the medicinal properties of 171 plants indigenous to India. It consists of three chapters. In Chapter 1, we have outlined 30 medicinal plant parts (bark, flowers, fruit, leaves, rhizomes, roots and seeds) with their scientific and commonly used names, provided chemical structures of some of the bioactive compounds found in these plants, and listed their medical uses in an easy-to-read bulleted format. We have presented essential oils separately in Chapter 2. The use of essential oils for therapeutic purposes is becoming increasingly popular. These oils remain in the human body for a limited time and do not leave residual toxins. This is in sharp contrast to chemical drugs, and their metabolic products, that usually take several months to exit the biological system. The use of essential oils is also eco-friendly. For example, it is more desirable to use the essential oil of oregano that is claimed to be about 30 times more potent than the conventional antiseptic, phenol.

In Chapter 3, we have listed the biological properties of 171 medicinal plants indigenous to India using a color-coded master table. A total of 55 biological properties related to different systems of the human body are outlined in the master table for the 171 plants. The usefulness of the master table is akin to the periodic table of elements. It provides a listing of the properties exhibited by different plants and gives a quick overview of different medicinal plants that exhibit the same type of activity. Thus, the master table can serve as a handy reference and save time and effort for scientists interested in researching medicinal plants. An appendix provides botanical names of the Indian medicinal plants outlined in Chapter 3. A list of sources and bibliography is provided at the end of the book.

We thank Dr. Piyush Pathak for typing the first draft of the manuscript, Tianyu She and Edward Barba for formatting the manuscript, and Professor Surinder Tikoo and Pankaj Dhar for reviewing the manuscript and making valuable suggestions. We also thank Mrs. Rupa Dhar, Mrs. Sonali Dhar, Nikhil Tikoo, Parth Dhar, and Pranava Dhar for their cooperation and understanding during the writing of this book. Without their support, this book would not have been completed.

Last but not the least, we are especially indebted to Ms. Sandhya Devi of World Scientific Publishing for the wonderful editorial support that she provided during the manuscript preparation and proofreading process.

Durga Nath Dhar
Retired Professor of Chemistry
Indian Institute of Technology, Kanpur, India

Preeti Dhar
Professor & Director, Biochemistry Program
Department of Chemistry
State University of New York
New Paltz, NY 12528 USA

Dedicated to the Memory
of
PANDIT SHAMBOO NATH DHAR
TULSI JEE KOTHIDAR
&
GOWRI JEE DHAR

Contents

Chapter 1

Thirty Medicinal Plant Parts and Their Medical Applications

APPLES: (*Malus domestica*), Pingguǒ (Chinese), Pommes (French), Apfel (German), Seib (Indian, Hindi), Mela (Italian), Manzana (Spanish).

> ➤ **Antidiabetic and cholesterol lowering activity:** Eating apples regulates the glucose concentration of blood and lowers cholesterol, due to the presence of soluble fiber (pectin).

> ➤ **Anti-hypertension:** The presence of potassium and phosphorus and associated diuretic effect of apples causes excess sodium ions to be expelled from the body with consequent lowering of blood pressure.

➢ Additionally, apples are beneficial for patients suffering from anemia and constipation.

ALOE VERA: (*Aloe barbadensis*), Lúhui (Chinese), Ghee kanvar (Indian, Hindi).

➢ **Moisturizer, analgesic, and healing activities:** The gel that oozes from a cut leaf is an excellent moisturizer; it is particularly useful for healing of burns caused thermally, by sunburn, or excessive exposure to X-rays. It provides a protective barrier against germs and hence minimizes irritation, providing immediate relief, and promotes quicker healing.

　　The mucilage obtained by breaking open the leaf of Aloe vera has an analgesic and healing effect. Nursing mothers can benefit by applying the gel on their sore nipples to achieve quicker healing and relief.

➢ **Anthelmintic, antiwrinkle, and anticonstipation:** The gel is reported to cure constipation and is associated with antiwrinkle and anthelmintic activities as well.

➢ **Helps in preventing the development of AIDS:** Acemannan, a bioactive molecule of Aloe vera gel is reported to stimulate the human immune system. It exerts synergistic effect on HIV drugs (AZT, ACV) and blocks HIV from developing into AIDS. With the addition of Aloe vera only 1/10 of the usual dose of AZT is needed for treatment, which reduces the cost as well as side effects of these drugs.

3' Azido 2', 3'
dideoxy thymidine
(AZT)

Acyclovir (ACV)

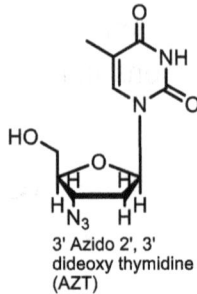

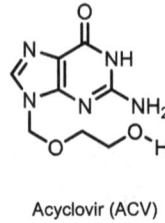

ANISEED: (*Pimpinella anisum*), Bājiǎo (Chinese), Anis (French), Anislikör (German), Velaiti saunf (Indian, Hindi), Semi d'anice (Italian), Anís (Spanish).

➤ **Mouth freshener:** The volatile oil in aniseed has an aromatic odor. Hence chewing these seeds refreshes the mouth.

➤ **Flatulence:** Aniseed tea removes flatulence and alleviates stomach disorders. The tea is prepared by adding a teaspoon of freshly ground aniseed in one cup of boiling water. The tea is allowed to simmer for five minutes and then filtered. It should be had slowly before meals thrice a day.

➤ **Laxative:** Aniseed exhibits a mild laxative action.

➤ **Asthma, cough, and bronchitis:** Aniseed has been found to be useful in treating asthma, cough, and bronchitis. The phytochemicals like cresol and α-pinene loosen bronchial secretions and make it easier to expel these secretions from the system.

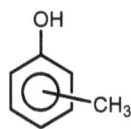

Cresol

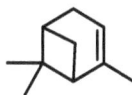

α-Pinene

➤ **Relieves menopausal symptoms:** Oral administration of aniseed is claimed to relieve the typical menopausal symptoms in middle-aged women. These symptoms arise due to the cessation of the production of female hormone-estrogen. Two constituents of aniseed, dianethole and photoanethole are phytoestrogens and mimic the role of the naturally occurring estrogen and thus provide relief to the menopausal patient.

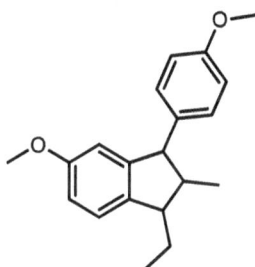

Dianethole

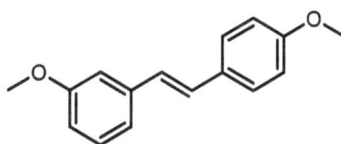

Photoanethole

➤ **Insecticidal properties:** The essential oil obtained from aniseed is reported to have insecticidal properties against head lice and mites that cause scabies. The major components of the aniseed essential oil are trans-anethole, γ-himachalene, trans-pseudo isoeugenyl 2-methyl butyrate, p-anisaldehyde, and methyl chavicol.

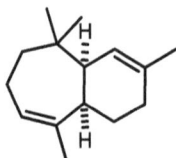

γ-Himachalene

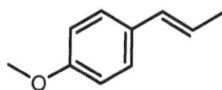

trans-Anethole

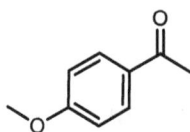

p – Anisaldehyde

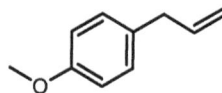

Methyl Chavicol

BASIL: (*Ocimum sanctum, Ocimum tenuiflorum*), Luólè (Chinese), Basilica (French, Italian) Basilikum (German), Tulsi (Indian, Hindi), Albahaca (Spanish).

> ➢ The leaves are pleasant smelling due to the presence of volatile terpenic oils (containing eugenol — 70%, eugenol methyl ether — 20%, and carvacrol — 3%), flavonoids, saponins, alkaloids, tannins, organic acids and vitamin C. These constituents find use in the treatment of skin disorders (acne)*, sore eyes, earache, headache, cough, and cold. The decoction of basil leaves also serves as a mouthwash and nasal douche.

> ➢ The leaves have antioxidant properties, relieve stress, are carminative, spasmolytic, antipyretic, control hepatic infection, and serve as an anthelmintic.

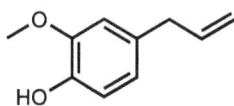

Eugenol

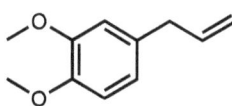

Eugenol methyl ether

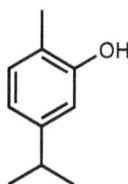

Carvacrol

> ➤ The leaves are associated with two important bioactivities: they are adaptogenic and have immune-modulatory actions.

* Acne- A skin condition that can be cured by the topical application of basil leaf paste on face, neck, upper part of chest and back. Alternatively, one can apply an infusion on skin. To prepare an infusion, about three teaspoons of dried basil leaves are added to a cup of boiling water and allowed to steep for about 15 minutes. The contents are allowed to cool, and the decoction applied to the acne.

> ➤ **Seeds:** The seeds have the following characteristics: mucilaginous, demulcent, diaphoretic, diuretic and employed in the treatment of cough, constipation, gonorrhea, diarrhea, piles, kidney disorders, and fever. They also help in relieving post child-birth pain.

CALENDULA: (*Calendula officinalis*), African marigold, Jīn zhǎn cǎo (Chinese), Zergul (Indian, Hindi), Calendula (Italian, Spanish).

> ➢ The phytochemicals present in Calendula plant are essential oils, salicylic acid, and a bitter substance called Calendulin. Calendulin exhibits antibiotic, antiseptic, and antifungal properties. British physicians exploited the antibiotic properties of Calendulin during World War I by preparing a dressing with a few drops of its alcoholic extract. The lotion of the flowers was also used as a wash liquid after surgical operations.

> ➢ Calendula is particularly useful for the treatment of chilblain –a skin condition due to inflammation of blood vessels from sudden warming from cold temperature and is characterized by severe itching and burning sensation.

> ➢ The infusion of the bright orange flowers of calendula applied over varicose veins and chronic ulcers, provides relief and ensures quick recovery.

> ➢ Calendula tincture is useful as a mouthwash, for ulcers, in particular, used after tooth extraction. Also used as an eyewash for conjunctivitis to cleanse and reduce the swelling of the eye.

> ➢ It is used for providing soothing relief from pain arising due to bruises and insect bites.

> ➢ The powder of leaves, used as a snuff, can induce the discharge of mucous.

> ➢ Calendula is bitter tonic, diaphoretic, antiemetic, and anthelmintic.

CAROM SEEDS: (*Trachyspermum ammi, Carum copticum*), Carom Zhŏngzĭ (Chinese), Grains de Carambola (French), Ajowan-Samen (German), Ajwain (Indian, Hindi), Semi de Carambola (Italian), Semillas de carom (Spanish).

➤ The principal constituent of the essential oil obtained from carom seeds is thymol (35–60%). Seeds also comprise carvacrol (also present in basil), p-cymene, β-pinene, limonene, β and γ terpinenes giving carom seeds their characteristic smell.

➤ The seeds are commonly used as a spice in curries. The phytochemicals present in the seeds help relieve stomach disorders and therefore chewing a teaspoon of seeds is a common household remedy for stomach disorders. The oral administration of seeds strengthens stomach function and gives relief from flatulence. A paste of the seeds is applied externally to lower abdomen region to relieve colic pains.

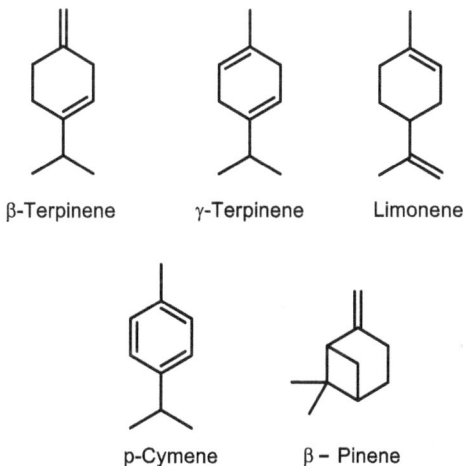

β-Terpinene   γ-Terpinene   Limonene

p-Cymene   β – Pinene

➤ The seeds in large doses should be avoided by pregnant women as they stimulate the uterus and can cause abortion.

➤ Other uses include the use of fruits for the treatment of asthma. Dry and hot fomentation of the crushed fruits on the chest helps relieve asthma symptoms.

➤ The tea may be prepared by putting freshly crushed carom seeds (one teaspoon) in two cups of boiling water. Water should be boiled till the volume is reduced to half. Tea can be consumed after cooling and straining the liquid. The tea is found to relieve acidity.

CHAMOMILE: (*Matricaria chamomilla*), Huáng Chūnjú (Chinese), Camomille (French), Kamille (German), Babuna (Indian, Hindi), Camomilla (Italian), Manzanilla (Spanish).

The steam volatile oil of chamomile plants is deep blue in color and the chemical constituents present therein are chamazulene, α (-) -bisabolol and its corresponding oxide.

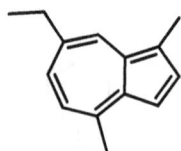
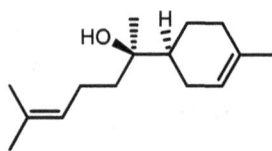

Chamazulene   α-(-) Bisabolol

➤ The oil finds various therapeutic applications (both as herbal and homeopathic medicine). It is carminative, spasmolytic, expectorant, anti-inflammatory, diuretic, anti-stress, CNS depressant, and sedative/tranquilizer. It is used for preventing sleep disorders.

➢ Promotes wound healing and can be applied, as a cream or a gel, on sore gums or on cracked nipples of nursing mothers.

➢ Useful in the treatment of backache, neuralgia, and arthritis.

➢ The flavonoids and the oil are useful in the treatment of skin disorders like psoriasis, eczema, and sunburn.

➢ Helps with teething problems in babies.

➢ Antibacterial, antiseptic, disinfectant, fungicidal and can be used as an ointment.

➢ Oil is used as a flavoring agent for alcoholic/non-alcoholic drinks, hair dye (the coloring imparted to the hair does not cause contact dermatitis), in the preparation of toothpaste, medicinal soap, body oil, and shampoo.

➢ +Antistress (tea): Two teaspoonfuls of dried chamomile herb can be placed in a cup of boiling water and steeped for about 10 minutes. Straining the liquid and consuming it once a day promotes a relaxing sleep and should be consumed before retiring to bed.

EUCALYPTUS: (*Eucalyptus globulus*), Ānshù (Chinese), Blue Gum Tree (Indian, Hindi), Eucalipto (Italian, Spanish).

The steam distillation of eucalyptus leaves and twigs yields a steam volatile oil, which contains a high proportion of citronellal (60–85%), citronellol, geraniol, and pinene.

> ➤ Inhalation of vapors derived from leaves gives relief from cough, helps loosen phlegm (so that it is easier to expel it), and clears stuffy nose.

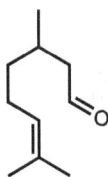 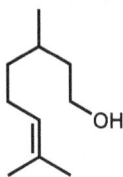 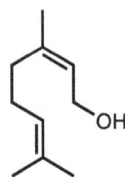

Citronellal   Citronellol   Geraniol

> ➤ Eucalyptus is anti-inflammatory and is used as an analgesic, sedative, carminative, febrifuge, expectorant, diaphoretic, diuretic, antiseptic, disinfectant, antiviral, antibiotic (citriodorol), insect repellant (citronellal) and provides relief in common cold, sore throat, whooping cough, fungal infections (athelete's foot), and eczema.

> ➤ Exerts deodorizing effect (namely when eucalyptus leaves are chewed, these perfume the breath).

> ➤ Hardens spongy and bleeding gums.

> ➤ Other uses: For scenting soap, detergents and room spray.

CINNAMON: (*Cinnamomum zeylanicum*), Ròuguì (Chinese), Cannelle (French), Zimt (German), Dalchini (Indian, Hindi), Canella (Italian), Canela (Spanish).

The inner bark/leaves of the cinnamon tree on steam distillation yield an oil which consists of eugenol and cinnamaldehyde.

The leaf oil has strong camphor like odor. The herb is used for several clinical applications (*vide infra*).

➤ The pungent odor of cinnamon promotes watery secretions from nose and respiratory tract, thereby clearing clogged passages. The patient suffering from influenza benefits by taking cinnamon tea.

➤ The tea is prepared by adding 4 grams of cinnamon powder to one cup of boiling water. Tea is allowed to stand for 15 minutes, and then strained. Several cups of the resulting decoction can be consumed in a day.

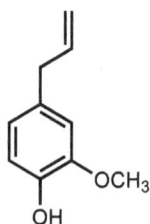

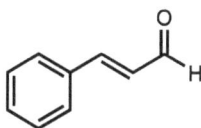

Eugenol   Cinnamaldehyde

➤ Cinnamon oil is antibacterial and is antiviral- in an indirect manner, viz; the viruses (bacteriophages) use the bacteria in the human body for their own multiplication, so when bacteria get killed by phytochemicals present in cinnamon, reproduction of virus gets automatically arrested and they get wiped out. The useful therapeutic properties of cinnamon are: anthelmintic, antidiahorreal, antidote to poisons, antispasmolytic, aphrodisiac, astringent, carminative, digestive, emmenagogue, hemostatic, refrigerant, and stimulant (circulatory, cardiac and respiratory).

➤ The other uses of cinnamon oil are for flavoring food, alcoholic, and soft drinks. Since both paste and oil of cinnamon are associated with fragrance, they are employed in the preparation of cosmetics, nasal sprays, as breath freshener, toothpaste, gum, and skin care products. Also employed as a liniment for rheumatic, head, and tooth aches. In powder form, it is used for correcting tedious labor, arising due to defective uterine contractions. Egyptians used cinnamon in preserving mummies based on their knowledge that cinnamon slows down putrefaction and decay.

TEA: (*Camellia theifera*), Chá (Chinese), Thé (French), Tee (German), Chai (Indian, Hindi), Tè (Italian, Spanish).

There are two main varieties of tea, viz; green and black. Green variety is obtained when freshly cut tea leaves are quickly dried and fired. Leaves that are fermented for several hours followed by drying and then firing, yield black variety. The volatile matter is present in green tea, while the non-volatile matter is present in black tea.

➤ Some phytochemicals in tea include, caffeine, theophylline, theobromine, tannic acid, gallic acid, and quercetin.

➤ The therapeutic effects associated with tea are that it is a stimulant, diuretic, and an astringent. When tea is boiled with water, for about ten minutes, caffeine, theophylline, and theobromine (CTT) are extracted but on prolonged extraction tannins start accumulating.

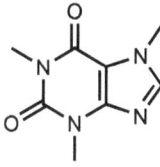 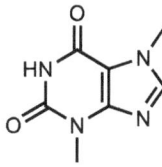 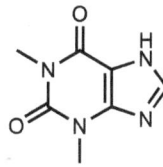

Caffeine   Theobromine   Theophylline

➢ CTT exert stimulating and refreshing action on biological system, while the presence of polyphenolics
(tannins) interferes with the digestive system.

➢ Tea serves as a nervine stimulant and has been found
useful in headache, neuralgia, and nervous depression. Tannin tea is useful for gargles.

➢ Based on astringent property of tea, it is employed in
the treatment of eye disorders. Tea extracts draw out
fluid from the swollen infected tissues and thereby
reduce swelling. Otherwise the swollen tissue would
serve home to bacteria and provide favorable conditions for their multiplication. For this purpose, the
tea leaves are boiled for some time, allowed to cool.
The liquid decoction is used to bathe eyes several
times a day.

➢ Some tea leaves are scented with the flowers of olive
or jasmine to impart fragrance to the drink.

COMMON STINGING NETTLE: (*Urtica dioica*),
Chángjiàn De cì Xún Má (Chinese), Ortie commune (French),
Brennesse (German), Bichu (Indian, Hindi), Ortica Commune
(Italian), Ortiga Común (Spanish).

➤ Stinging nettle is reported to contain a high concentration of chlorophyll, iron, and vitamin C. The leaves of this plant have a skin irritant effect (*vide infra*).

➤ Though a troublesome weed, it is medicinally a useful herb. It is claimed to provide moderate relief from allergy symptoms like stuffy nose, watery eyes, and hay fever.

➤ Since the herb is rich in iron and vitamin C, it is known to elevate glutathione levels in healthy red blood cells.

➤ It also lowers blood sugar levels and hence is potentially useful as an anti-diabetic drug. The recommended dose of medication is one teaspoonful of nettle suspended in 250 ml of warm water. After waiting for about 15 minutes, the decoction is consumed.

➤ Contact with nettle leaves results in nettle rash due to the presence of formic acid in the stings. Interestingly, the tincture (alcoholic extract) prepared from the herb gives relief from nettle rash and other eruptive problems.

The following therapeutic properties are associated with the herb:

➤ It exerts a powerful hemostatic action and has been found to be useful for treating bronchial and uterine hemorrhage.

➤ Steam cooked young leaves of nettle have a strong laxative effect.

➤ The inhalation of dry powder has been found to provide relief to patients suffering from asthma or bronchial problems.

➤ The pulverized dry herb checks nasal bleeding when used as a snuff.

➤ Additionally, the herb has been used as an astringent, a diuretic, an anthelmintic, and emmenagogue.

JASMINE: (*Jasmine grandiflorum*), Mòlì (Chinese), Jasmin (French, German), Chameli (Indian, Hindi), Gelsomino (Italian), Jazmín (Spanish).

➤ The fragrant flowers of jasmine are reported to calm the nerves and aid in stomach problems.

➤ Jasmine is a conglomerate of about 100 pleasant smelling constituents viz; benzyl alcohols, fornesol, cis-jasmone, linalool, methyl anthranilate, methyl jasmonate etc.

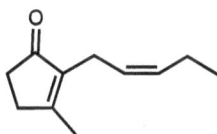

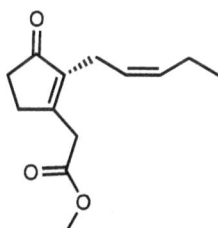

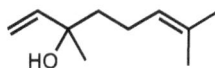

cis-Jasmone   Methyl-Jasmonate   Linalool

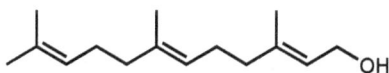

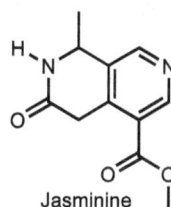

Farnesol   Jasminine

> ➤ The leaves exert a mild analgesic effect and are reported to contain salicylic acid and an alkaloid, 'Jasminine'. The whole plant is associated with therapeutic properties like- antidepressant, anti-inflammatory, antiseptic, carminative, diuretic, expectorant, and hair tonic.

> ➤ Jasmine oil exerts a cooling effect when applied externally to patients suffering from headache or skin diseases.

> ➤ Leaf juice exhibits a remedial effect on ulcerations in mouth, gum, or throat.

> ➤ Massaging with jasmine is stimulating and facilitates childbirth.

> ➤ Arouses sexual desire when the flowers are rubbed around the genital area.

> ➤ Jasmine is used as an important ingredient in the manufacture of soaps, cosmetics, and perfumes.

> ➤ It is used in the food industry for imparting fragrance to alcoholic and soft drinks.

> ➤ Dried flowers are used for preparing jasmine tea.

BLACK MUSTARD: (*Brassica nigra*), Hēi Jièmò (Chinese), Moutarde Noire (French), Schwarzer Senf (German), Sarson (Indian, Hindi), Senape Near (Italian), Mostaza Negra (Spanish).

> ➤ It is used internally as a condiment and externally as an oil or paste for the treatment of several medical

conditions. Caution should however be exercised in its use, as it may cause strong reaction including blisters, on skin and mucous membranes.

➢ The seed contains fixed oil (20–25%) as pale-yellow liquid associated with a sharp penetrating odor but lacks any essential oil. Allyl isothiocyanate is generated, when the seed comes in contact with water (during steam distillation).

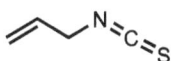

Allyl Isothiocyanate

Some important medicinal applications of mustard are:

➢ For stopping obstinate vomiting: The poultice or plaster from mustard flowers is applied on the upper abdominal region.

➢ The difficulty in breathing that occurs in whooping cough is eased by the application of mustard paste.

➢ It has been reported to be effective for treating swollen glands, headaches, cerebral congestion, cardiac and chest pains.

➢ When used judiciously, mustard seeds alleviate the pain arising from arthritis, sciatica, and backache.

➢ Mustard oil, diluted with rubbing alcohol alleviates pain when applied to any painful area. Mustard plaster may be prepared as follows: Prepare a thick paste by mixing mustard powder with cold water and spread it over a clean cloth. The mustard covered cloth is covered by a layer of gauze, before being transferred to the designated area (say, on the back) for a total period not exceeding 20 minutes. Strict observance of these instructions will prevent skin irritation and burns in the patient.

➤ Additionally, mustard is antimicrobial, antiseptic, diuretic, emetic, febrifuge, and rebefacient.

BARBERRY: (*Berberis aristata*), Fú Niú (Chinese), Berberitze (German), Rasaut (Indian, Hindi), Crespino (Italian), Bérbero (Spanish).

➤ The chief constituent of barberry is the alkaloid berberine, which is antibacterial and antiviral.

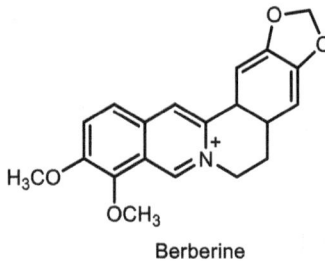

Berberine

➤ It stimulates the immune system and activates white blood cells to kill and devour invading microorganisms. The effect can be achieved by consuming one cup of barberry tea per day.

➤ The tea can be prepared by putting half teaspoon of the powdered root bark into a cup of boiling water and allowing it to stand for about half an hour. After cooling the decoction, honey can be added to offset the bitter taste (due to berberine) and consumed.

➢ Eye infection can be remedied by using barberry as a compress. A clean piece of cloth dipped in barberry tea can serve as a compress.

➢ The common name of the extract from barberry (or related plants of the barberry species) is called 'Rasaut'. It is employed as a purgative for children. Rasaut is antipyretic, astringent, helps to reduce bleeding in piles, diaphoretic, diuretic, cures jaundice, helps relieve painful urination, has stimulant action on the gastro-intestinal tract movement, relieves vomiting during pregnancy, and serves as an excellent thirst quencher.

PEPPERMINT: (*Mentha piperita*) Bòhé (Chinese), Menthe Poivrée (French), Pfefferminze (German), Pudina (Indian, Hindi), Menta piperita (Italian), Menta (Spanish).

The yield of essential oil obtained by steam distillation of flowering herb is about 4%. It is a pale-yellow liquid with a penetrating camphor like odor. The principal constituents of the herb are menthol, (~20%), menthone, 1,8 cineol, limonene, menthyl acetate and pulgeone.

➢ The high menthol content imparts the oil the characteristic cooling and refreshing properties.

➢ It is an excellent appetizer and improves digestion.

➢ The peppermint tea is prepared by placing 4–5 mint leaves in cup of boiling water and letting the water stand (covered with a lid) for about 15 minutes. The decoction (tea) is strained and consumed twice in a day (before lunch and after dinner).

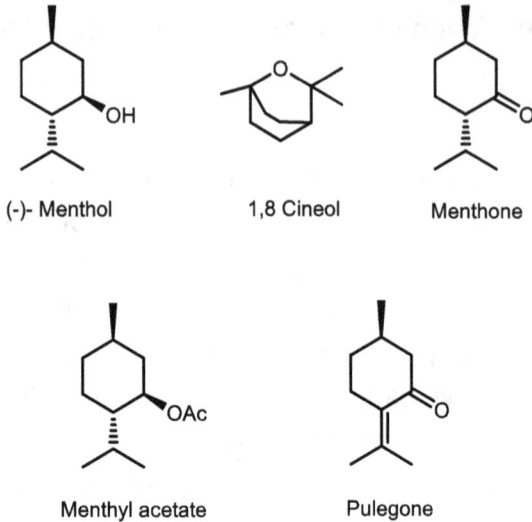

(-)- Menthol   1,8 Cineol   Menthone

Menthyl acetate   Pulegone

> ➤ The tea provides relief from disorders of stomach, namely, bad breath, dyspepsia, flatulence, gastric spasm, hyperacidity, nausea and vomiting.

> ➤ The tea is reported to be refreshing and alleviates stress related problems like anxiety, hysteria, and insomnia.

> ➤ The patient suffering from skin disorders are benefitted by taking a hot water bath containing the peppermint herb and it is known to invigorate the bather.

> ➤ The herb acts like a magic wand in alleviating pains of various types. For example, pain arising due to headache, migraine, ear and tooth aches (caused by caries) and menstrual pain.

Other health promoting applications of the peppermint herb are its antiseptic, deodorant, stimulant, carminative, antispasmodic* and antiemetic properties.

* The combined effect of antispasmodic and 'cooling' effect makes peppermint herb an excellent inhaler for asthmatic patients.

Other applications are:

➤ Flavoring agent in pharmaceuticals (used in cold and digestive remedies).

➤ Flavoring for food (Chewing gum, soft and alcoholic drinks, candy).

➤ Fragrant constituent in soaps, tooth pastes, detergents, cosmetics and perfumes.
Largely used in pharmaceutical preparations to disguise the taste of unpleasant smelling drugs.

➤ Keeps ants and mites at bay.

PASSION FLOWER: (*Passiflora foetida*) Jīqíng huā (Chinese), Fleur de la passion (French), Passionsblume (German), Jhumkalata (Indian, Hindi), Fiore Della Passione (Italian), Flor de la pasión (Spanish).

This herb is widely planted in India.

➤ The plant is the one of nature's best sedatives.

➤ The dried leaves and stem contain bioactive molecules such as maltol, ethyl maltol, alkaloids, and flavonoids. Some of these constituents relieve muscle and nervous tension and as a consequence exert favorable effects on the following health conditions: Insomnia, nervous tension, muscle spasm, fatigue, and depression.

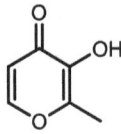
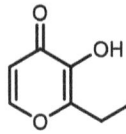

Maltol   Ethyl Maltol

➢ Flower extracts have exhibited an interesting effect on hyperactive children; flips mood and aids in their mental concentration.

➢ Decoction of the leaves has also been found useful in biliousness and asthma.

➢ Leaves are used on the head to control giddiness and give relief from headache.

➢ Fruit is reported to be emetic and contains HCN.

➢ Due to the presence of bioactive molecules in leaves and flowers of passiflora, their extracts are widely used in medical practice as nerve calming agents.

➢ Employed as a raw material in the production of soft drinks due to its fragrance and mucilagous nature.

ST. JOHNS' WORT: (*Hypericum perforatum*), Shèng yuēhàn de màiyá zhī (Chinese), Millepertuis (French), Johanniskraut (German), Bassant (Indian, Hindi), L'erba di San Giovanni (Italian), Hierba de San Juan (Spanish).

➢ The flowers are known to have a tranquilizing effect and the calming action lowers the anxiety and depression (particularly in women suffering from emotional disorder related to menstrual cycle*).

* Consuming tea prepared from the flowers of St. Johns' Wort is found to be useful in correcting this disorder. The dried plant (one tea spoon) is put in 500 ml cold water and the water brought to a boil, followed by cooling to room temperature. The decoction is consumed in gastric disorder and for regulating menstruation.

➢ The herb stimulates the gastrointestinal secretions particularly from liver (bile).

➢ It helps in healing through lesser scarring of scalds caused by blisters, cuts and minor wounds.

➢ It prevents hemorrhagic tendency and is therefore given to patients recovering from surgery. When applied over a wound, the red pigment- hypericin (in flowers) in combination with other antibiotics present (including flavonoids) suppress the spread of infection and inflammation and expedite the healing process. The leaves of the herb have the following additional therapeutic uses: anthelmintic, antidiarrhea, diuretic, emmenagogue, purgative, spasmolytic and vermifuge.

➢ Leaves are used as a cure for prolapsed uterus/anus.

ONION: (*Allium cepa*), Yángcōng (Chinese), Oignon (French) Zwiebel (German), Pyyaz (Indian, Hindi), Cipolla (Italian), Cebolla (Spanish).

Steam distillation of onion yields a pale-yellow oil, associated with a strong unpleasant odor and lachrymatory effect.

The principal constituents of the onion bulb are organo-sulphur compounds (diisopropyl disulphide, allyl propyl disulfide, dipropyl trisulfide) and catechol and protocatechuic acid.

The therapeutic uses of onion are many. The following few examples are illustrative:

> Their consumption raises the 'good cholesterol' (HDL) in blood.

> It possesses blood thinning property and does not allow cholesterol to get deposited within the arteries. Since cholesterol deposition in arteries could lead to a heart attack, eating raw onions (half raw onion/day) can prevent heart attacks.

> Toothache can be alleviated by putting a small piece of onion on the effected tooth or gum. Heated onion juice can be used as an ear drop to remove earache.

> People suffering from protruding piles get a lot of relief if a compress of roasted onion bulbs are applied over it.

> A liniment of onion juice has been found to provide relief when applied over painful joints or swollen part of skin.

> Other useful therapeutic applications of onions are: anthelmintic, antimicrobial, antirheumatic, antiseptic, antispasmodic, antiviral, bactericidal, carminative, diuretic, expectorant, emmenagogue, hypoglycemic,

hypotensive, rubefacient, stomachic, tonic for hair growth, vermifuge.

GURMAR: (*Gymnema sylvestre*) Gourmar (French).

The leaves (and roots) of Gurmar contain a mixture of triterpene saponins, trace amounts of alkaloid-gymnamine and gymnemic acid, among others.

The plant is indigenous to India. It has been shown to have antidiabetic property. Thus, after taking the herb orally the patient suffering from type I diabetes would require lesser amount of Insulin injection. While in the case of type II diabetics, sugar concentration in the blood may drop to the level that they may not need further medication. In sharp contrast to synthetic oral antidiabetic drugs, the herb will lower the blood sugar levels only if it is high (compared to standard value) and will not lower it to dangerously low levels. According to indigenous system of medicine, equal quantities of air dried powder of leaves of gurmar and giloy (*Tinospora cordifolia*) are consumed for effective treatment of diabetes.

Other therapeutic applications of the herb are:

> ➢ The application of leaf paste over the eyes has been found useful in curing corneal opacity.

> ➢ The formation of dental plaque is prevented due to the presence of gymnemic acid in the herb.

> ➢ Gastric disorders get corrected by consuming the leaf juice, twice/day.

> The whole plant is used in the treatment of tuberculosis.

> For controlling malarial fever, the patient is orally administered a filtered aqueous extract of the finely ground leaves. The recommended dose is 2–3 teaspoon of the extract per day for three days.

> Patients suffering from urinary troubles benefit by consuming 100 ml of extract of the leaves twice a day.

> Recently the herb has been used for body weight reduction.

> Additionally, the leaf and roots of gurmar are reported to exhibit the following effects: antiperiodic, astringent, diuretic, laxative, refrigerant, and stomachic. Also used as a tonic.

PSYLLIUM: (*Plantago ovata*), Chē qián zi (Chinese), Isabgol (Indian, Hindi).

The constituents are: pentosans, aldobionic acid and linoleic acid.

> The therapeutic uses are the following: antidiabetic, anti-inflammatory, prevents arteriosclerosis, demulcent, emollient, and a laxative. Ingestion of 10 g of psyllium/day for a month is claimed to increase HDL and reduce serum cholesterol by 9.6 % and triglycerides by 8.6%.

> Swelling and pain resulting from rheumatism and gout can be remedied by applying a poultice made of psyllium seeds, oil, and vinegar.

➢ Aqueous decoction of psyllium, mixed with honey is a good treatment for sore throat and bronchitis.

➢ It relieves abdominal pain.

➢ Good for diabetes control. Relieves polyuria.

➢ Checks diarrhea.

➢ The mucilage of psyllium lubricates the ulcerated surface of inner walls of intestinal mucosa. Removes toxins from the guts, protects the intestinal walls from irritation by certain foods. It also cures chronic dysentery (including slimy dysentery) and eliminates the *Entemoeba histolytica* from the intestines.

➢ It corrects constipation. Psyllium absorbs large quantities of water, and the mucilageous material produced therefrom serves two purposes. It lubricates and promotes bowel contraction and increases the bulk of fecal matter and thus aids in its expulsion from the body.

➢ Other uses: As a stabilizer in chocolate manufacture, cosmetic formulation, ice cream industry, textile sizing, and preparation of commercial gum.

SAGE: (*Salvia officinalis*), Zhìzhě (Chinese), Sauge (French), Salbei (German), Salvia (Indian, Hindi), Saggio (Italian), Sabio (Spanish).

Principal constituents: borneol, camphor, caryophyllene, cineol (p-16), pinene (p-3,7), thujone (42%), bitter principals, resin and tannins.

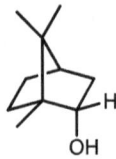

Borneol

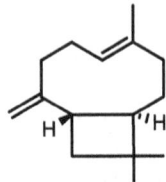

Caryophyllene

Camphor

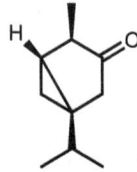

Thujone

➢ Sage is specific for inflammation of mouth, gums, tongue, throat, and pharynx. Gargles with the decoction of sage (aerial parts and seeds) is useful due to inflammatory, astringent, and antiseptic properties that arise due to the presence of tannins.

➢ It is powerfully nervine and acts on the cortex of brain, thereby improves memory, concentration, and eliminates mental exhaustion.

➢ Eliminates involuntary trembling of the limbs.

➢ It lowers blood sugar level in diabetics. For this, a decoction of the dried herb (1–2 teaspoon of dried leaves) is prepared in boiling water and let stand for about 10 minutes. It is consumed (1 cup) early in the morning.

➢ De-stress: To neutralize the effect of stress or our system, it is advisable to consume the tea prepared either from dried sage or fresh leaves in the usual way.

➤ Sage has been reported to exert remedial action on menstrual disorders, checks the tendency of habitual abortion, and is used in the treatment of chronic liver diseases.

➤ The therapeutic applications of the herb are: anti-diabetic, antidiarrheal, anti-inflammatory, anti-microbial, antioxidant, antiperspirant, antiseptic, astringent, blood purifier, digestive, diuretic, antihy-pertensive, laxative, moth repellant, stimulant (bilary secretion), and a tonic.

➤ It is employed in the manufacture of mouthwash, tooth paste, fragrant component of soaps and shampoos.

➤ The oil and oleoresin from sage are used extensively for flavoring foods, soft drinks, and alcoholic beverages.

➤ The leaves are used in salads and sandwiches and it is a popular herb in culinary preparations.

ROMAN CARAWAY: (*Carum carvi*), Luómǎ xiāngcài (Chinese), Carvi romain (French), Römischer Kümmel (German), Shah Jeera (Indian, Hindi), Cumino (Italian), Alcaravea (Spanish).

The dried fruit is crushed, and steam distilled. The volatile oil is a pale-yellow liquid, which turns dark on keeping. The organic constituents of caraway herb are as follows: caryo-phyllene (p-21), cuminaldehyde (~60%), cymene, farnesene, carvone, limonene, myrcene, phellandrene, pinenes, terpinenes.

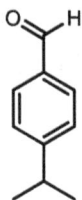
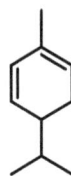

p-Cymene   Cuminaldehyde   Phellandrene

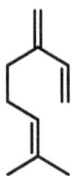
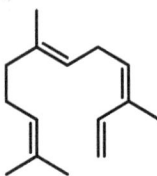

Myrcene   Farnesene

> ➤ The herb is anthelmintic, antidiarrheal, antihistamine, antioxidant, antiseptic, aromatic, astringent, carminative, diuretic, emmenagogue, expectorant, larvicide, laxative, spasmolytic, stimulant, tonic, and vermifuge.

> ➤ Externally, it is employed as a poultice to allay pain and irritation of worms in the stomach. Internally, it can be consumed shortly after childbirth since it promotes secretion of milk in the mother. The poultice prepared from crushed seeds has been reported to reduce the swelling of breasts and testicles.

> ➤ The herb is useful in restoring normal voice after having a rough/harsh voice.

> ➤ Provides relief from insomnia, renal colic, common cold, and amnesia.

> ➤ The oil is employed as a flavoring ingredient in pharmaceutical products to mask the bitter or obnoxious odor of medicines.

> As a fragrance component in toothpaste, cosmetics, and perfumes, as well as a flavor to many foods.

TURMERIC: (*Curcuma longa*), Jiānghuáng (Chinese), Safran des Indes (French), Kurkuma (German) Haldi (rhizome) (Indian, Hindi), Curcuma (Italian, Spanish).

Contains curcumin, dihydrocurcumin, zingiberene, tumerone, borneol, sabinene, phellandrene, among others. The rhizomes are cleaned and sundried. The material is then steam distilled to yield an essential oil, which is a yellow orange liquid with a bluish fluorescence and a spicy odor.

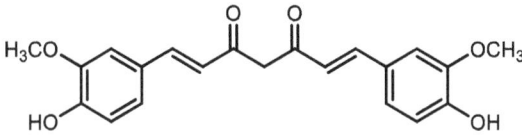

Curcumin

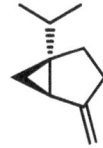

Sabinene

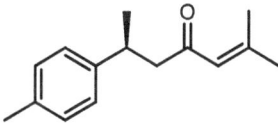

ar–Tumerone

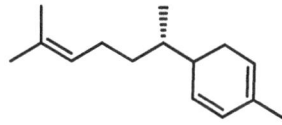

Zingiberene

> The therapeutic uses are: analgesic, anthelmintic, antiarthritic antibacterial, anti-diarrhoeal, antiflatulent, antifungal, anti-inflammatory, anti-oxidant, aromatic, carminative, digestive, diuretic, heart and liver protective, antihypertensive, insecticidal, laxative, stimulant, styptic, and a tonic.

Following are some useful recipes for healing several types of health disorders:

> For eye problems, decoction of turmeric can be used as an eye wash.

> For sprains, bruises, and inflammatory problems of joints, a paste of turmeric is made by mixing it with quicklime (or alum). The paste has been reported to promote scab formation over the eruptions produced in smallpox and chickenpox.

> For treating purulent discharge resulting from nasal catarrh, application of a mixture of turmeric powder and powdered alum (1:20) has been found to be beneficial. Patients suffering from nasal catarrh benefit by inhaling the fumes arising from burning turmeric.

> The important application of turmeric is for controlling the formation of intestinal gas. This is because the population of gas producing bacteria is diminished substantially by the antibiotic action of the rhizome.

> The major component of turmeric is curcumin, which is anti-inflammatory. It is recommended to take one teaspoon of turmeric in one cup of warm milk, thrice a day.

> Regular intake of turmeric has shown to improve dementia and Alzheimer disease symptoms.

> In culinary practice, it is an indispensable ingredient in some foods, meat dishes, and condiments. It imparts a

yellow color to food and is also used in confectionery and pharmaceutical industries.

FENNEL: (*Foeniculum vulgare*), Huíxiāng (Chinese), Fenouil (French), Fenchel (German), Sauf (Indian, Hindi), Finocchio (Italian), Hinojo (Spanish).

The fennel fruit yields 3–5% of the volatile oil of pleasant aromatic odor and exerts carminative and antispasmodic action. Principal constituents are anethole (50–60%), limonene (p-7), phellandrene (p-23), pinene (p-3), anisic acid and anisic aldehyde.

➤ It exhibits the following therapeutic properties: anthelmintic, anti-inflammatory, antimicrobial, anti-rheumatic, antiseptic, aphrodasiac, appetizer, carminative, depurative, diuretic, emmenagogue, expectorant, galactagogue, laxative, purgative, stimulant, splenic, stomachic, tonic, and vermifuge.

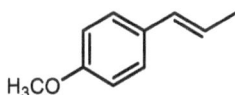

Anethole

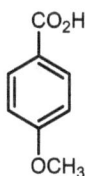

p-Anisic acid

➤ Administration of fennel improves eyesight as well as memory.

➤ A mixture of fennel and the cardiotonic herb (Lily of the valley) is claimed to cure stroke.

- ➤ It benefits blood circulation, muscles, and joints.

- ➤ Fennel exerts a beneficial effect on edema, obesity, and rheumatism.

- ➤ Brushing teeth with crushed fennel seeds and baking soda eliminates bad breath.

- ➤ The oral administration of fennel seeds eliminates flatulence and gets rid of catarrhal and phlegm from the bronchial tubes of the lungs.

- ➤ Fennel water (~100 ml) is a reputed digestive aid for flatulence, constipation, and dyspepsia. Chewing the seeds after meals not only prevents bad breath but eliminates the possibility of vomiting and indigestion.

- ➤ The tea prepared from fennel seed is described to be specific remedy for relieving colic pain in a baby. The tea is prepared by taking teaspoon of crushed fennel seeds and boiling in 250 ml of water for about 20 minutes. The decoction is filtered, cooled and given to the baby.

- ➤ It sets the painful menstrual irregularities in order. Additionally, fennel has proved useful in increasing the milk output of nursing mothers. These effects have been rationalized due to the presence of estrogenic components in the herb.

- ➤ Fennel is incorporated in some therapeutic products viz; carminatives or laxative preparations; to impart flavor to the medication; or to soft and alcoholic drinks.

➤ Because of its good smell, fennel oil forms a constituent of soaps, toiletries, and room sprays.

FENUGREEK: (*Trigonella foenum-graecum*), Hú lú bā (Chinese), Fenugrec (French), Bockshornklee (German), Methi (Indian, Hindi), Fieno greca (Italian), Fenogreco (Spanish).

Constituents are diosgenin, trigonelline, trigocoumarin, yamogenin, mono-, di and trimethylamines. The herb contains fairly large quantity of organo-iron.

➤ The following properties are associated with the use of fenugreek: antidiabetic, anti-diarrhoea, anti-dysentery, anti-dyspepsia, anti-flatulence, anti-gout, anti-inflammatory, counters enlargement of liver and spleen, demulcent, diuretic, emmenagogue, emollient, hair tonic, lactogogue, nutritive, and a tonic.

➤ Fenugreek tea is reported to be an aid for indigestion, flatulence, and sluggish liver. It is prepared by placing 2 teaspoons of the herb in one cup containing hot water; let stand for about 5 minutes and filter. The decoction thus obtained may be sweetened with honey (and lemon). Dosage: one cup per day.

➤ The tea derived from fenugreek seeds has the advantage of being used as a gargle for sore throat. Used also for ulcers of stomach and intestines. Due to the mucilaginous nature of fenugreek, the irritation as well as the ulcers get cured in the long run. The presence of estrogenic component in the herb is reported to be useful in the treatment of cancer. Also, it stimulates the secretion of milk in nursing mothers.

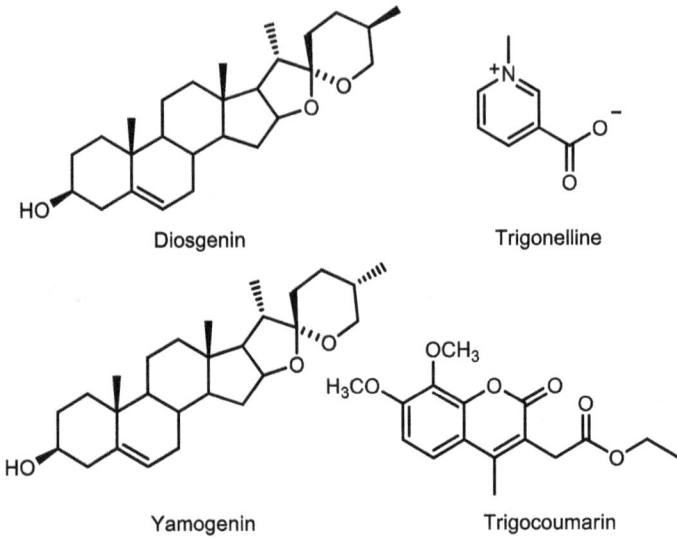

Diosgenin

Trigonelline

Yamogenin

Trigocoumarin

> ➤ The most important property of fenugreek is its antidiabetic action. One can take two teaspoons of fenugreek powder (seeds) and ingest it with milk or water. It is reported to lower the levels of blood sugar, cholesterol, and triglycerides.

> ➤ One important property of the herb is that it exerts a cleaning action in human body and restores the deadened senses of taste and smell.

> ➤ Bad odors coming out of mouth or body arise due to the accumulation of toxins in the system. Fenugreek expels the malodorous components of our body and thus helps keep the body healthy and clean.

> ➤ Flour of the seeds serves as good poultice due to 30% mucilaginous content and is used as a cosmetic

(on skin), applied to head to prevent hair loss, including promoting hair growth, and for bringing boils and carbuncles to mature.

PARSLEY: (*Petroselinum crispum*), Xiāngcài (Chinese), Persil (French), Passionsblume (German), Prajmoda (Indian, Hindi), Prezzemolo (Italian), Perejil (Spanish).

Parsley seeds contain a glucoside, apiin, and a pale-yellow essential oil, the main constituent of which is apiol. Additionally, it contains myristicin, tetramethoxyallyl benzene, pinene, flavonoids, and volatile fatty acids, including vitamin A and ascorbic acid (in leaves).

➢ The therapeutic uses of parsley are: antimicrobial, antipyretic, antiseptic, astringent, carminative, depurative, diuretic, emmenogogue, expectorant, febrifuge, laxative, stimulant, stomachic, and tonic.

➢ Nursing mother can suppress the secretion of milk, by applying the leaves over the breasts, several times a day.

➢ Bruised leaves of parsley are used as poultice for sore eyes. Parsley helps in reducing the inflammation of the delicate membrane of the inner surface of eyelids and corrects the opacity of the lens of the eye.

➢ Parsley leaves contain a high content of chlorophyll and vitamins A and C. Chewing parsley leaves, acts as a mouth freshener.

➢ Digestive disorders are accompanied by the formation of gas within the stomach and intestines. Oral

administration of four teaspoons of herb with a glass
of water provides relief to the patient.

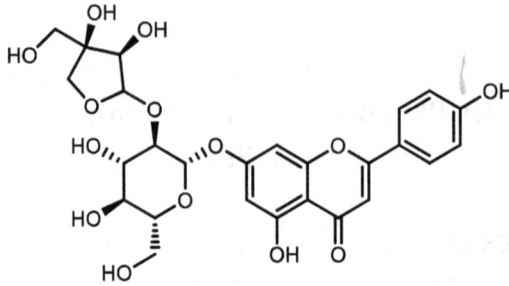

Apiin

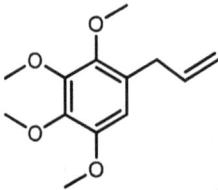

Tetramethoxyallyl benzene

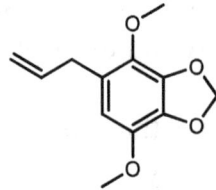

Apiol

> Apiol present in the herb, acts like an estrogen
  (female sex hormone). Oral intake of the herb is
  reported to be effective remedy for various types of
  menstrual disorders (example, the monthly periods
  that are either absent/scanty and painful). Parsley is
  also found useful to relieve cramps.

> The seeds contain a volatile component, which is
  claimed to promote blood circulation in organs
  located in the pelvic region of our body.

> Parsley exerts a tonic effect on the urinary system.
  viz; kidney and bladder. Additionally, it prevents the

formation of urinary calculi (in kidneys and bladder). Parsley has been reported to be an effective remedy for dropsy.

➤ Due to its aromatic flavor, it finds application as a culinary herb in the preparation of soups and other dishes.

➤ The seed oil is used as a constituent in soaps, detergents, cosmetics, and perfumes. It finds extensive use as a food flavoring agent (including alcoholic and soft drinks).

VALERIAN: (*Valeriana officinalis*), Xié căo (Chinese), Valériane (French), Baldrian (German), Jalakan (Indian, Hindi), Valeriana (Italian, Spanish).

The essential oil, obtained by steam distillation of roots and rhizomes of this plant is an olive-brown colored liquid with a musty odor. Constituents include valepotriates, a volatile oil consisting of esters of valeric and isovaleric acids, borneol, patchouli alcohol, pinene, valerianal, valeranone and two alkaloids chitinine and valerianine.

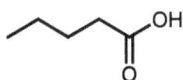

Valeric acid

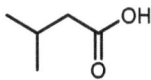

Isovaleric acid

Borneol

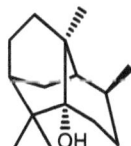

Patchouli alcohol

➤ Therapeutic action of valerian is as follows: anodyne (mild), antidandruff, antispasmodic, bactericidal, carminative, CNS depressant, diuretic, hypnotic, hypotensive, regulator, sedative, spastic, stimulant, and stomachic.

➤ Some of the manifestations of nervous disorder (listed next) are amenable to treatment with the oral administration of valerian herb: Migraine, vascular excitation, restlessness, irritability, nervous tension, epilepsy and hysteria (victims are women of all ages). An efficient and safe treatment, for the above disorders consists of taking about 30g of the fresh herb and infusing it into 500 ml of water, and consuming the extract, four times a day. The prescription is also a cure for insomnia. It has the advantage that it calms the mind without inducing drowsiness in the patient.

➤ The herb has been reported to exhibit a curious effect on some animals (cats, dogs, horses, rats, and mice). Thus, cats look jubilant and frisky after smelling the herb. Similar behavior has been observed in horses, rats, and mice. The unfriendly dogs are brought to submission by gypsies by feeding them with food containing a small amount of oil derived from valerian and aniseed. Likewise, the scent of valerian has been exploited as bait to trap small animals like mice and rats.

➤ The oil of valerian is used in pharmaceutical preparations, as a relaxant.

➤ It is also employed as a fragrant component in soaps and provides flavor to roots liqueur.

GARLIC: (*Allium sativum*), Dàsuàn (Chinese), Ail (French), Knoblauch (German), Lahsun (Indian, Hindi), Aglio (Italian), Ajo (Spanish).

Essential oil of garlic is obtained by steam distillation of fresh crushed bulbs. It is a pale-yellow mobile liquid with a strong unpleasant odor. Principal constituents are allicin, ajoene, allyl propyl disulphide, diallyldisulphide, diallyl tri-suphide, citral, geraniol, linalool and phellandrene.

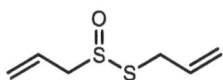
Allicin

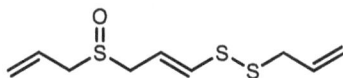
E-Ajoene

➤ The therapeutic properties of garlic are: amoebicidal, anthelmintic, antiseptic, antimicrobial, antibiotic, antitumor, antiviral, aphrodisiac, bactericidal, carminative, cholagogue, depurative, detoxifying, diaphoretic, expectorant, febrifuge, fungicidal, hypoglycemic, hypotensive, insecticidal, larvicidal, promotes leukocytosis, stomachic, and tonic.

➤ **Earache:** Placement of a small clove of garlic in the ear of a patient is claimed to be the wonder cure for earache. The sulphur containing antibiotic allicin, present in garlic, kills the causative agent and as a consequence brings relief to the patient. Alternatively, warm garlic oil can be introduced into the ear and covered with a wad of cotton. To obtain garlic oil, the garlic clove is mashed in olive oil and left standing at ambient temperature for about 4 days. The oil is strained and kept in refrigerator and when required, a small amount of oil is warmed and applied repeatedly to the ear at a four-hour interval.

➢ **Lung disorders:** Research has shown that garlic has a wonderful capacity to combat infection caused by bacteria, fungi and, viruses. For example, bronchitis, produced by one of these microbes, is curable by ingesting several cloves of garlic in a day. The other lung disorders that get relief from eating garlic include pulmonary tuberculosis, whooping cough, cough, common cold, and bronchial congestion. Garlic when given in sufficient doses is an invaluable remedy in the treatment of pneumonia. Due to its antimicrobial activity, garlic is useful in the treatment of diarrhoea, dysentery, and gastro-intestinal infections. Garlic also helps control the symptoms of cystinosis, a lysosomal storage disease.

➢ **Diabetes:** Taking a few cloves of garlic, remarkably reduces the blood glucose level, in humans, to normal levels. As a bonus it reduces the risk of heart disease and protects one from infections.

➢ **Heart diseases:** Three health benefits accrue by taking three cloves of garlic (upto a maximum of 10) in a day. By virtue of its anticoagulant components, it is an excellent blood thinner; and puts the blood circulation in a top gear, thus prevents stroke, decreases blood pressure, and reduces cholesterol level to normal. It may be noted that cardiovascular benefits arise due to the presence of allicin and ajoene in the garlic bulbs.

➢ **Skin disorder** (cuts, scrapes, bruises, and wounds): A paste is prepared by mashing a clove of garlic that is then applied on the area having minor skin infections. It is bandaged and let stand overnight. In this way the infective organism is eliminated, and healing process takes place. The antibiotic power exerted by garlic is comparable to the well-known antibiotic penicillin.

➢ Rubbing garlic over ringworm provides relief. Mustard oil or coconut oil in which garlic has been fried is an excellent antiseptic for ulcerated surface wounds, maggot infested ulcers, and scabies.

➢ **Athlete's Foot:** Athlete's foot, a fungal disease, is curable by garlic. Thus, rubbing garlic over the area, and in addition it is recommended to consume one or two cloves every day. Following the procedure ensures that the infection will not recur.

➢ Digestive disorders: Garlic has antiflatulent action and exerts a stimulating effect.

➢ Garlic is extensively used as a spice and for flavoring. It is employed as an ingredient in many health food products designed to reduce high blood pressure and protect from heart disorders. Incorporation of garlic in food products imparts a characteristic taste and flavor.

GINGER: (*Zingiber officinale*), Shēngjiāng (Chinese), Gingembre (French), Ingwer (German), Adrukh (Indian, Hindi), Zenzero (Italian), Jengibre (Spanish).

Ginger on steam distillation yields a pale-yellow viscous oil with the yield ranging from 1–5%. It has a characteristic aromatic odor. The chemical constituents present in ginger are: borneol (p-29), camphene, cineol, phellandrene (p-23) and zingiberene. The non-volatile oil contains the yellow and pungent component of ginger.

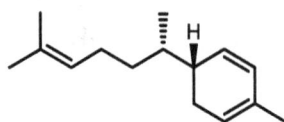

Zingiberene   Cineol

> **Uses:** the pharmacological properties of ginger are: antiemetic, antihistamine, anti-inflammatory, aphrodisiac, carminative, diuretic, expectorant, febrifuge, rubefacient, and stalogogue.

> Ginger finds application in the treatment of broad range of ailments: Aches and pains, arthritis, bronchitis; cardiac disorder, colic, common cold, diarrhea, dropsy, indigestion, menstrual disorders, neuralgia, rheumatism, skin burns, tuberculosis (pulmonary), vertigo, and whooping cough.

> **Arthritis:** Ginger is a reputed herbal remedy for treating nausea and motion sickness. In rheumatic arthritis, oral administration of ginger (1 g powder or 5 g of fresh material) is reported to give substantial relief compared to the conventional drugs, with

respect to the overall improvement in the symptoms. This includes, for example, decrease in swelling, greater flexibility of the joints, and less pain.

➢ **Skin burns:** Ginger has been found to heal skin burns and reduce associated inflammation. The herb accelerates the repair of the damaged tissues and dulls the stinging sensation. Hence fresh ginger root juice or crushed ginger can be applied over the burn. One can also use it to get relief from sunburn.

➢ **Common cold:** Common cold is usually associated with fever, pain in the body, and excessive perspiration. A useful recipe consists of taking 2–3 g of a mixture of ginger, black pepper, long pepper, licorice (all in equal quantities), plus a few leaves of tulsi, 2–3 times per day. To make the cough productive, fresh grated ginger with honey should be consumed. Gives quicker relief from cough because of its expectoration property.

➢ **Indigestion:** Stomach disorder is indicated when patient has a feeling of heaviness, gas formation, and loose motion or constipation. The recommended recipe for correcting this condition is by taking 2–3 times a day, a decoction made by adding 5 g of dried ginger powder to 1 liter water and bringing it to a boil. For controlling diarrhea, 1–3 g of dried ginger is sweetened with equal amount of sugar/honey and is taken orally, twice a day.

➢ **Burns on skin:** Burns caused by hot water or fire can be relieved by applying a small piece of cotton, moistened with ginger extract. Ginger is a blood thinner, hence its oral administration improves blood circulation. It also prevents atheriosclerosis.

➢ **Aches and Pains:** Aches and pains are alleviated by applying ginger (paste or extract) at the site of pain. Thus, headache can be relieved by applying the paste of dry ginger (in water) on the forehead. Likewise, toothache is relieved by rubbing ginger paste on the gum. For allaying earache a few drops of ginger should be instilled into the problem ear.

➢ **Other uses:** The essential oil derived from the rhizome of ginger is used in the manufacture of flavoring essences and in perfumery. It is widely used for culinary purposes on account of its aromatic and pleasant odor. Ginger is also used in the preparation of condiments, curries, and gingerbread.

LICORICE: (*Glycyrrhiza glabra*), Gāncǎo (Chinese), Réglisse (French), Lakritze (German), Mulethi (Indian, Hindi), Liquirizia (Italian), Regaliz (Spanish).

The chief constituents of licorice are: Glycyrrhizen (6–10%), which is 50 times sweeter than sugar and has been identified as a tri-terpenoidal saponin, together with glycyrrhic acid and glycyrrhetimic acid, glucose (4%), sucrose (2.5–6.5%), glycyramarin-the bitter principal, flavonoidal glycoside-liquritoside, asparagin (2–4%), volatile oil and fat. The therapeutic uses of licorice are listed as alterative, demulcent, emollient, expectorant, laxative and pectoral.

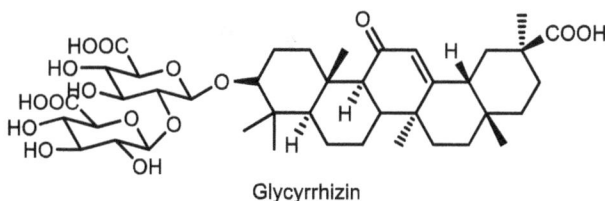

Glycyrrhizin

> **Arthritis:** The bioactive molecule present in licorice is 18β- glycyrrhetinic acid, which is claimed to be pain relieving and also exerts an anti-inflammatory action. Licorice also exhibits anti-rheumatic-arthritic activity. This capability is the result of stimulation in the production of two hormones (cortisone and aldosterone). In this connection it is advised to take two cups of licorice-tea in a day, which can be prepared by steeping (for 10 minutes), half teaspoon of the powdered herb in a cup of boiling water.

> **Skin disorders:** The bioactive molecule present in licorice acts like cortisone, and hence has proved beneficial in the treatment of skin disorders, like, dermatitis, eczema, and psoriasis. It is important to note that the like of cortisone (occurring naturally in licorice) is completely free from adverse effects associated with the synthetic cortisone.

> **Respiratory disorders:** Licorice is endowed with anti-bacterial, antiviral, antiallergic, and anti-inflammatory properties. This has been ascribed to be due to the presence of glycyrrhizin in the herb. Licorice is an effective remedy for cough, colds, flu, and sore throat. Licorice has antibacterial properties and has the capability of

killing streptococci bacteria, which are responsible for sore throat problems. Additionally, it numbs the pain, and brings about a soothing effect on sore throats. The in-built expectoration property in licorice ensures the removal of the bacterial debris and morbid sputum from the respiratory tract. For general protection licorice-tea should be taken at the rate of 2 cups per day (method of preparation, *vide supra*).

➢ **Anti-gastric activity:** Liquiritoside, the flavonoid component of licorice, possesses anti-spasmodic activity. The presence of additional bio-active compounds, carbenoxolone, prevents gastric juice from attacking the stomach lining. On both these counts licorice finds application for treatment of stomach ulcers.

➢ **Rejuvenating action:** Licorice has rejuvenating and aphrodisiac action. French women use it as a secret preparation for enhancing their love life.

The herb has the following additional attributes:

➢ It quenches thirst.

➢ Remedies feverishness.

➢ Corrects the inflammation and irritation in the bronchial tubes, including intestine and catarrh of the gastro-urinary passages.

➢ Licorice is commonly used as a flavoring component in beverages and confectionary. It is used in pharmacy in masking the bitter and acrid taste of nauseous drugs, particularly aloes and senna (leaves).

Caution: The use of licorice is not recommended for pregnant or nursing mothers, including patients having diabetes, glaucoma, heart diseases, hypertension, or history of stroke.

Chapter 2

Essential Oils and Their Therapeutic Uses

2.1 Introduction

Aromatherapy involves the use of essential oils to improve the quality of life at many levels (physical, emotional, and spiritual). These essential oils tend to calm, balance, and rejuvenate body, mind, and spirit. Essential oils are extracted from plants and work in harmony with our body. As an adjunct to modern medicine, the use of essential oils, as a therapeutic aid, has gained public interest worldwide since 1992. People have discovered the beneficial effects and the aesthetic enjoyment derived from the use of essential oils and this has resulted in their widespread use. Some essential oils (e.g. tea tree oil) have good antiseptic properties. Essential oil derived from oregano has 26 times more powerful antiseptic action than phenol, a principal constituent of commercially available antiseptic formulations. The yield and the purity of the essential oil depends upon various parameters. Yield depends on parameters such as genetic make-up, geographical location, soil, climatic conditions and harvesting time whereas purity is a function of the methods adopted for extraction from the plant material.

2.2 Occurrence

The following parts of some species of plants bear the aroma bearing liquids: bark, flowers, fruits, leaves, roots, seeds, wood, needles, and twigs.

2.3 Aromatherapy

The term aromatherapy was introduced in 1928 by a French chemist (Rene-Maurice Gattefosse´) who had a perfumery business. Gattefosse´ received severe burn injuries on his hand while working in the laboratory and out of sheer desperation, applied some lavender oil to the burn injuries. He noticed that the pain in his hand subsided and several of the burn marks healed quickly without leaving any scars.

Essential oils by and large are non-invasive and non-toxic to human body and have broad spectrum applications. They are constituents of medicine, food, and cosmetics. The bitter taste of some modern medicines can be masked by incorporating essential oils, which impart a natural flavor and fragrance to the preparation. These oils are reported to possess cell rejuvenating and beautifying properties. They also have soothing aroma and emotion enhancing properties. Aromatherapy is used in combination with other therapies to treat illness and alleviate symptoms. There are about 300 essential oils in general use today and typical essential oil bearing plants are:

- Chamomile
- Clove
- Eucalyptus

- Geranuim
- Lavender
- Lemon
- Peppermint
- Rosemary
- Tea tree
- Thyme

Essential oils have broad spectrum applications and have been used as antibacterial, antiviral, antifungal, analgesic, antidepressant, antineuralgic, antispasmodic, nervine, sedative, anti-inflammatory, and granulation tissue stimulant. Some of essential oils are used to ease digestive problems. Additionally, essential oils are reported to promote blood circulation, which in turn helps oxygen and nutrients reach each cell in the body, enhancing the elimination of waste products from the system and thereby increasing the immunity of the body.

2.3.1 Chamomile: (*Matricaria chamomilla*), Huáng Chūnjú (Chinese), Camomille (French), Kamille (German), Babuna (Indian, Hindi), Camomilla (Italian), Manzanilla (Spanish).

> ➢ Chamomile (German/Roman) is associated with antiseptic, disinfectant, and anti-inflammatory action.

> ➢ Useful for skin condition (eczema, psoriasis), treatment of fevers, nervousness and depression.

> ➢ The valuable property of chamomile is that it is a mild sedative and exerts a calming action.

2.3.2 Clove: (*Syzygium aromaticum*), Dīngxiāng (Chinese), Clou de girofle (French), Nelke (German), Laung (Indian, Hindi), Chiodo di garofano (Italian), Clavo (Spanish)

> ➤ It is well known that clove oil can cure toothache. This is the consequence of its antibacterial and analgesic action. However, concentrated oil should not be used on gums/skin. In earlier times, clove oil was used for the sterilization of surgical instruments because of its strong antiseptic action.

> ➤ It possesses sedative property.

> ➤ Clove oil helps with digestion and muscular disorders. When taken orally, it helps overcome the feeling of nausea. Using clove oil provides relief to asthma patients.

2.3.3 Eucalyptus: (*Eucalyptus globulus*), Ānshù (Chinese), Neelgiri (Indian, Hindi), Eucalipto (Italian, Spanish).

> ➤ Commonly used in the treatment of cough and common cold.

> ➤ Oral use of the essential oil is supposed to cool our body in the summer and provide protection during winter.

> ➤ Other important applications exploit anti-inflammatory and antibiotic properties of the oil. Eucalyptus essential oil has been used for skin related problems and as an analgesic.

2.3.4 Geranium: (*Pelargonium graveolens*), Tiānzhúkuí (Chinese), Géranium (French), Geranie (German), Geranium (Indian, Hindi), Geranio (Italian, Spanish).

> ➤ One of the outstanding properties of Geranium oil is its usefulness in skincare. For example, claims have been made that the oil from this plant provides relief from a condition called chilblain.

> ➤ Geranium oil is reported to serve as a nerve tonic as well as a relaxant and a sedative.

> ➤ It is generally found useful for alleviating symptoms resulting from disorders of menopause, throat, blood and serum glucose.

2.3.5 Lavender: (*Lavandula officinalis*), Xūnyīcǎo (Chinese), Lavande (French), Lavendel (German), Laivendar (Indian, Hindi), Lavanda (Italian, Spanish).

> ➤ It is a versatile and indispensable essential oil that can be put to many uses. Lavender is known to be effective treatment for burn and scald injuries. On healing, it does not leave any white patch (sear) on the skin. It is a natural antiseptic.

> ➤ Lavender acts as a sedative and antidepressant.

> ➤ It stimulates the immune system of our body.

> ➤ Lavender is a mood elevator and helps the patient bear the psychological shock resulting from an injury.

2.3.6 Lemon: (*Citrus limonum*), Níngméng (Chinese), Citron (French), Zitrone (German), Neemboo (Indian, Hindi), Limone (Italian), Limón (Spanish)

> ➢ Lemon oil is antiseptic and hence is useful as a water purifier.

> ➢ It serves as an antidote for some insect venoms.

> ➢ Lemon oil is a remedy for tension headache.

> ➢ It serves as a stimulant for digestive system and a tonic to the lymphatic system.

> ➢ It is claimed to help reduce body weight.

2.3.7 Peppermint: (*Mentha piperita*), Bòhé (Chinese), Menthe poivrée (French), Pfefferminze (German), Pudeena (Indian, Hindi), Menta (Italian), Menta (Spanish).

> ➢ From ancient times, peppermint oil has been used effectively in reparation of disorders related to digestion and blood circulation. It helps with flatulence, indigestion, flu, migraine, toothache, and rheumatism.

> ➢ It acts as repellant for rodents (mice) and insects (fleas and ants).

> ➢ Mint tea is supposed to lower blood pressure. This tea can be prepared by adding 5–6 mint leaves to a cup of boiling water. The tea can be consumed after the mint leaves have been in water for about 5–10 minutes.

2.3.8 Rosemary: (*Rosmarinus officinalis*), Mí dié xiāng (Chinese), Romarin (French), Rosmarin (German), Rojamairee (Indian, Hindi), Rosmarino (Italian), Romero (Spanish).

> ➤ Rosemary oil is used as a component of morning bath as it serves as a mental stimulant. It is known for relaxing the muscular tension resulting from long physical activity and hence rosemary bath at the end of the day is relaxing and rejuvenating.

> ➤ The versatility of its application can be gauged by its usefulness in different kind of conditions like depression, migraine, headache, memory loss, arthritis, diabetes, cough, and flu.

2.3.9 Tea tree: (*Melaleuca alternifolia*), Cháshù (Chinese), Théier (French), Tee Baum (German), Albero del tè (Italian), Árbol de té (Spanish).

> ➤ The tree indigenous to Australia has a broad range of antibiotic (antibacterial, antiviral and antifungal) activities. It is claimed to be one hundred times more effective than phenol, a known antiseptic.

2.3.10 Thyme: (*Thymus vulgaris*), Bǎilǐxiāng (Chinese), Thym (French), Thymian (German), Banajwain (Indian, Hindi), Timo (Italian), Tomillo (Spanish).

Thyme oil should be used in moderation and not applied to the skin without dilution with a suitable carrier oil.

➢ It has been used for respiratory infections like flu. An aromatherapy diffuser can be used to disperse thyme oil, including the carrier oil, in the air of the room inhabited by the patient, to eliminate the virus.

➢ Thyme finds application in treatment of a wide range of conditions, like acne, warts, whooping cough, rheumatism, and neuralgia.

➢ Thyme oil is a strong repellant and prevents parasites and insects from entering homes.

2.4 Early History

In olden times, Chinese utilized plant aromatics for therapeutic purposes and for performing some religious rituals. Examples of plants used as aromatics and for religious ceremonies by Chinese are ginger, camphor, and opium. Egyptians developed perfumes and burnt incense in temples. They also seemed to have nurtured the art of cosmetics for beautification purposes. Egyptians had known about the scrubs for embalming the dead with herbal preparations. The knowledge about the use of herbs spread to other parts of world, particularly to Israel following the mass migration of Jews from Egypt to Israel. Composition of anointing 'oils' for priesthood was a closely guarded secret but it was subsequently revealed to be consisting of cassia, cinnamon, myrrh, and olive oil. Additionally, when westerners became aware of the usefulness of herbs, the oriental merchants seized the opportunity of introducing the herbal treasure of the East to the West. Examples are: camphor (China), cinnamon (India) and rose (Syria). Eventually, pure odoriferous oils were obtained by distillation.

It was the Greek physician Hippocrates, who recommended the use of pleasant smelling oils as perfumes. He also recommended them as a remedy for wounds and for inflammation of the skin. Like Egyptians, Romans and Greeks also used fragrant oils in their rituals. Besides, Romans used fragrant oils for massaging purposes and in baths. Avicenna, the Arabian physician and scholar is credited with the discovery of distillation of essential oil from rose, involving the use of refrigerated coils.

2.5 Later Developments

Eventually, European perfumers successfully isolated fragrant oils from indigenously available herbs (such as lavender, rosemary, and sage). The isolation of essential oils was followed by attempted identification of major constituents of these oils. It was in this context that the following constituents were identified for the first time in the history of perfumes, viz, cineol, citronella, and geranial. It has been observed that eucalyptus oil has a stronger antiseptic action than its isolated principal constituent, cineol. Likewise, rose oil is a multicomponent system; the principal odoriferous constituent is phenyl ethyl alcohol. The sum total of all the constituents present in rose oil lead to a harmonious blend giving rose its characteristic smell. However, if we were to compare the smell of a natural rose with that of phenyl ethyl alcohol, the odor of the latter is no match with the pleasant odor of the natural rose oil.

A French doctor (Jean Valnet) was the first to successfully use essential oil in the treatment of a medical condition (psychiatric disorder). This was followed by the work of a French woman researcher who studied the rejuvenating properties of some oils and exploited these for 'beauty therapy'. She designed her aromatherapy formulation suited to the temperament of each individual person. Other than imparting a pleasant smell, essential oils are found to produce several health benefits. When essential oil is rubbed on the skin, it is immediately absorbed into the system. This is evidenced by the observation that when a clove of garlic is rubbed on the soles of feet, the volatile components of garlic diffuse through the skin and mingle with the circulating blood, and the characteristic smell of garlic appears in the breath, a little later. The effect of the oil operates at three different levels, viz; pharmacological, physiological, and psychological. The pharmacological effect is exerted by the interaction of the oil with enzymes and hormones present in the blood. The physiological effect is reflected by its calming or stimulating action. The psychological effect is the response of the person to odor after it is inhaled or rubbed.

Many medicinal herbs also yield the volatile essential oils. Both constituents (oil and the residual herb) have been reported to have therapeutic qualities, with little difference. For example, chamomile (Roman as well as German) yields an oil that has anti-inflammatory, antispasmodic, wound healing, skin soothing, and pain-relieving properties. Additionally, it serves as a muscle and nerve relaxant, as it helps to relax tensed up muscles and is hence used to treat insomnia. The herb contains other constituents (not present in the volatile oil) such as tannins, bitter components, and mucilages. The intact medicinal plant contains small percentage of essential oil, in addition to the constituents mentioned above (tannins, mucilages, and bitter components). For therapeutic purposes, the herb is processed by either making an infusion (just made by steeping the plant part in water), decoction (a more concentrated form of infusion), or tincture (extraction usually done with alcohol). In this way (say, by infusion) the concentration of the volatile oil constituents extracted is less and its potency is reduced accordingly. The herbal preparation is best suited for internal therapeutic use, since it contains other useful constituents (*vide supra*) which, however are lacking in the essential oil obtained from it.

Another illustration is that of peppermint. The essential oil obtained from peppermint is reputed to be antispasmodic and antiseptic in action. Based on these properties, it is useful as an inhalant for respiratory disorders. It is advisable to extract the whole peppermint plant, if one is required to use it for a prolonged time. In that event, the volatile as well as non-volatile components will exert their combined therapeutic effects. In certain situations when better results are required, then it is advisable to combine peppermint with other appropriate herb(s) as well. For example, peppermint, cornmint, and spearmint contain one common constituent, menthol. Incidentally, all three herbs belong to Labietae family. Similarly, two herbs [chamomile (both Roman and German variety) and yarrow] belonging to the compositae family of plants contain a blue colored volatile oil, called chamazulene.

2.5.1 Caution in the use of essential oils

Caution should be exercised when using essential oils. Undiluted essential oil when applied on skin, or used in aromatherapy, can bring about irritation and damage. It is fortunate that these volatile oils are soluble in non-volatile oils called carrier oils. Essential oils therefore should be diluted in carrier oils and then applied safely onto the skin. The volatile components of the essential oils from the diluted oil, when inhaled, travel into our respiratory system without any difficulty. It should be kept in mind that volatile oils (essential oils) are obtained in concentrated form. Their dispensing is therefore made in drops rather than in gram quantities. It is worthwhile to remember that healing property is not the sole monopoly of essential oils. A case in point is that of plantain herb, which has wound healing property, although the herb does not contain any essential oil.

Volatile oils derived from spearmint and cornmint have been used in small quantities for various purposes (such as imparting flavor to food and beverages (alcoholic as well as non-alcoholic), pharmaceuticals, cosmetics and as perfumery constituents in soaps, creams, lotions.)

While using the same volatile oil, attention should be paid to the following instructions with respect to their effect on the body:

- Undiluted volatile oil of say cinnamon should not be used. Using concentrated cinnamon oil obtained from the leaf is alright as long as it is used in moderation. The presence of eugenol in the oil irritates the mucous membranes; while the bark oil is not only an irritant to skin but makes it photosensitive and results in skin pigmentation.
- Some volatile oils are safe to use in the concentrated form (without dilution with a carrier oil). Examples are sandalwood, ylang ylang, and lavender oils.

- There is a long list of volatile oils that should be avoided, by either pregnant women, or those who suffer from medical conditions, like high blood pressure, epilepsy, diabetes, and neurological disorders.

2.5.2 Present status

Folklore has given us some medicinal herbs discovered by our ancestors by hit and trial method, spread over a period of several centuries. The plants identified through folklore having curative properties, have now been adopted for use by practitioners of modern medicine. Systematic analysis of the medicinal herbs has revealed that the curative property is due to the presence of biologically active compounds present in them (e.g. quinine). Quinine was first isolated from the bark of Cinchona tree and the bark of this tree has been used to treat malaria for centuries. The quest for discovering more plants associated with health benefits is in progress in various research laboratories worldwide. We have to bear in mind that many plants are, unfortunately becoming extinct and as a consequence, we may be losing a treasure of therapeutic information in the years ahead. In contrast, the research on volatile oils with respect to their health benefits is scanty.

It is interesting to note that the herbal plants and their action on the human body are described in the modern medical jargon as prophylatic, anti-depressant, and nervine tonic and are no different in meaning from what folklore had described them centuries ago. It turns out to be the question of different nomenclature being used to describe a human ailment. The following example is illustrative. Basil was described to provide protection against evil (prophylactic), the fragrance was good for the heart (cardiotonic) and removed soreness (anti-depressant, since relaxation techniques are also used to treat depression). Like herbal remedies, the volatile oils can exhibit stimulating action in one part of the human body, and yet at the same time produce a sedating effect elsewhere.

2.5.3 Skin

The problems associated with skin should not be seen in isolation. They actually arise due to some underlying factors such as accumulation of toxins in the blood, imbalance in the hormonal levels in the body, and also due to nervous and emotional troubles. Essential oils diluted with non-volatile oil, or with alcohol provide relief and restore the disordered condition of skin.

Lavender oil has useful applications in the following skin disorders:

- Insect Bites
- Eczema and Infected wounds
- Ringworm
- Burns
- Excessive Perspiration

2.5.4 Circulatory system

Essential oils when rubbed over the skin, diffuse into blood stream, thereby altering its circulation. There are some volatile oils that exhibit rubefacient effect. They dilate the blood vessels and bring in more blood in that area, reducing inflammation and providing considerable relief from pain. Volatile oil derived from Hyssop is reported to provide a balancing effect; viz; it reduces the blood pressure, if it is high, and raises it to normal value if the circulation of blood is sluggish and poor. (vide, adaptogens).

2.5.5 Respiratory system

Nose, throat, and lung infections are amendable to treatment with volatile oils. When inhaled through nose, the vapors pass through the bronchial tubes and then go to lungs. The bronchial tubes react with the oil, producing bronchial secretions, which are helpful in controlling the spread of respiratory diseases. The essential oils are absorbed through the skin into the blood stream faster as compared to when they are ingested. When taken orally, essential oils are disposed of mainly through the lungs, but a small amount is secreted through urine. Examples include volatile oils derived from eucalyptus, sandalwood, chamomile, frankincense, and thyme.

2.5.6 Digestive system

For correcting digestive ailments pertaining to stomach and liver, it is advisable to take the medicinal herb instead of volatile oil alone. The presence of tannins, bitter principles, and mucilaginous components also help with the digestive ailment. These components are however lacking in the volatile oil. Some herbs that aid in digestive problems are chamomile, cinnamon, lavender and peppermint.

2.5.7 Reproductive system

Some essential oils as stated earlier, penetrate through skin and mix with the blood stream. The volatile oil from jasmine and rose are reported to bring about hormonal changes which result in strengthening of the reproductive system whereby the problems related to sex and menstrual cycle are taken care of. The volatile oils from hops and fennel contain plant hormones and thus can be used for correcting the endocrine problems of females, viz; regulation of menstrual cycle and lactation. Volatile oils from herbs influence secretions from different glands. For example, rose, geranium, lemongrass, clove, spearmint, and peppermint affect thyroid secretions and cinnamon, clove, lavender, and rosemary affect adrenal gland secretions, including the generation of estrogen and the male sex hormones.

2.5.8 Immune system

Nature has provided us with white blood cells (WBCs) which engulf foreign germs invading the organism and devor them. Nearly all the essential oils exert a dual action. They not only promote the production of WBCs but are also lethal to the infecting agent. The net result is that the immunity of the individual is enhanced. Some volatile oils hence can contain the spread of infections due to malaria, plague, and typhoid. It is common observation that individual who use these oils regularly are less prone to diseases (common cold, for example) and if they do get infected, they recover rather quickly. The representative examples are, the volatile oils derived from lavender, clove, peppermint, and chamomile.

2.5.9 Nervous system

Nervous system responds differently to essential oils. Some oils exert a stimulating effect on the central nervous system while other set of oils, viz lavender, chamomile and sandalwood have a sedating effect. Some volatile oils such as geranium are adaptogens. Geranium sometimes behaves as a stimulant and at other times it exerts a sedative effect, depending on each individual.

2.6 Method of Application of Essential Oils

At home, essential oils can be safely used to serve as perfumes as well as first aid. Minor ailments like headache, toothache, body ache, and muscle cramps can be treated with appropriate essential oils. However, one should seek a doctor and not resort to aromatherapy for treatment of major ailments.

The volatile oils should be stored in dark airtight amber or blue colored bottles, away from the heat and with no direct light. This prevents photo-degradation and aerial oxidation of the oil.

2.6.1 Massage

Full body massage with an essential oil that suits the patient is recommended. The volatile oils must be diluted with a carrier oil before using it for massage (preferable carrier oil is sweet almond, grape seed, apricot, avocado, and jojoba oil). The concentration of the volatile oil in the blended oil should be to the extent of 3 percent. Nervous or emotional patients would require a lower concentration of blended oil compared to patients suffering from rheumatism. Massaging the oil gives a relaxing experience. The places that can be self-massaged are palms of hands and the soles of the feet. If for example, one is suffering from indigestion, then rubbing peppermint oil in the area of stomach in a clockwise manner is of help. Likewise, the stiffness in neck and shoulder muscles can be relieved by applying the volatile oil derived from marjoram. The massaging of the delicate areas viz; around the eyes should be done smoothly while exercising caution. The three principal steps involved in massaging are:

- Stroking (effleurage)
- Kneading (potrissage)
- Friction (circular movements)

For treating some skin problems such as herpes or athlete's foot, oregano and tea tree oil (about 6 drops) may be diluted with one teaspoonful of vodka or isopropyl alcohol. The resulting solution may in turn be diluted by the addition of 1 liter of distilled water. The diluted solution may subsequently be used for washing of open wounds or pits of genital herpes or chicken pox.

2.6.2 Hot and cold compresses

For alleviating pain or reducing inflammation on any part of the body, recourse is taken to aromatherapy wherein one uses hot and cold compresses. This is carried out in the following way: To a bowl of hot water, 5 drops of essential oil are added, and a cotton swab or handkerchief is dipped into it, and excess water taken up in the fabric is squeezed out.

The cotton fabric is then placed on the target site, till its temperature equals the body temperature of the patient and the cycle is repeated several times. Hot compresses are used for conditions like rheumatism, arthritis, and aches related to tooth, ear, or body. In the case of cold compresses, ice cold water is substituted for hot water. This procedure has been found useful for conditions like headache, backache, and inflammation.

2.6.3 Hair care

The best method of hair care consists of taking chamomile derived volatile oil (3%) in olive oil, fortified with sweet almond or jojoba oil. The oil so prepared is massaged into the scalp, followed by wrapping the hair in a warm towel for about an hour. The above-mentioned method, conditions the hair, and gets rid of parasites, like lice.

2.6.4 Flower water

Flower water is useful for some skin conditions like derma- titis and eczema and also for improving the complexion of the face. Flower water is prepared by adding about 30 drops of essential oil derived from rose into 100 ml of deionized water and let stand for several days in dark before filtering it. The filtrate is then used as described (*vide supra*).

2.6.5 Baths

Romans enjoyed the benefits derived from aromatic baths. Lavender and chamomile derived volatile oils have the added advantage of being useful for stress related relaxation and for treatment of insomnia. Likewise, pine and rosemary volatile oils are known to soothe tired and aching limbs.

2.6.6 Vaporization

In order to avoid the dust and smoke that arises by burning of incense sticks, one can vaporize an essential oil through the aromatherapy diffuser. Another option is to put a small wad of cotton, impregnated with one drop of essential oil, and placed on top of the electric bulb, which may be turned on when required to release the aroma molecules into the atmosphere. It induces a peaceful and relaxed atmosphere. To recreate this type of environment, frankincense and cedar-wood have traditionally been used in certain rituals. Similarly, the volatile oils derived from citronella and lemongrass are used to keep away mosquitoes and biting insects; or for clearing the air from cigarette smoke. To clear the atmosphere of harmful microorganisms, which often times are cause of epidemics, the use of essential oil (in vapor form) is a must. People suffering from breathing problems can be relieved by keeping the volatile oil of myrtle or eucalyptus in their bedrooms. A few drops of essential oil can be spread over pillow for night and on handkerchief for use throughout the day.

2.6.7 Douche

The common infection of the genito-urinary canal of females can be controlled by giving enema with warm water (1 liter) containing 5–6 drops of tea tree oil. To accelerate the healing process after childbirth, one can use douche containing lavender or cypress oil.

2.6.8 External application

Arthritis is the result of built up of toxin at the joints. Adherence to a restricted diet along with the use of herbal remedies and external application of volatile oils (derived from white birch or juniper), is necessary to get relief in arthritis.

2.7 Synergy

Some volatile oils blend well with each other, such that the perceived resultant effect is more pronounced than the numerical combination of two components. Thus, lavender and chamomile oils exert a synergic effect on each other. In general, the volatile oils of the same plant species blend. Thus, peppermint blends well with spearmint, since they both belong to Lamiaceae family and both contain menthol. Some blending combinations on the other hand, do not match with each other and exert an inhibiting power over one another.

2.8 Adaptogen

The dual action of an essential oil to bring about a balance in the human body is due to its adaptogenic behavior. Examples of this category include hyssop, lemon, peppermint, ginseng, and mint. Thus, the volatile oil derived from hyssop has the capability of reducing the high blood pressure of a patient to normal level or raising the low blood pressure to the normal level. Lemon for example can be used as a tonic or a sedative, depending upon the requirement of the situation. Peppermint behaves as both, a stimulant or a relaxant to establish a balance in the system.

2.9 Personal perfumes

It is an art to blend aromatic oil in such a manner that it suits to one's personality. For a beginner, experimentation with blending is the only way to learn about the finer details of the art.

2.10 Habitat

It is a well-known fact that the aromatic plants grown in different regions, under different growth condition and different climatic conditions produce differing components from each other. So, when describing a medicinal herb, it becomes necessary to specify these parameters in detail.

2.10.1 Composition

Following are the components found in various essential oils.

Terpenes: Camphene, caryophyllene, limonene (citrus oils)
Sesqui-terpenes: Chamazulene (chamomile oil)
Diterpene: Myrecene, phellandrene
Aldehydes: Citral: starting material for the synthesis of perfumery material-β-ionone.
Ketones: Camphor, carvone, jasmone from jasmine- eases congestion and aids free flow of mucous. Menthone, pulegone (source: pennyroyal), sage, thujone.
Esters: Borneol acetate, eugenol acetate, linalyl acetate (from lavender), used as fungicides and sedative
Alcohols: Benzyl alcohol, borneol, citronellol (source: rose, eucalyptus), farnesol, geraniol (source: palmarosa), linalool (source: lavender), turpeneol, vetiverol (good antiseptic properties and generally non-toxic to humans).

2.11 Methods of Extraction

Essential oils are generally extracted from flowers, seeds and roots. After harvesting, the flowers should be subjected to extraction right away, to avoid putrefaction of the material on storage. This applies to seeds or roots, which should be stored with caution until the time of their processing. The process of extraction consists of expression (application of pressure); examples are citrus oils, particularly the ones derived from lemon. The other methods of extraction are based on distillation in the presence of steam, water, or in dry conditions. The essentials oils from lavender, sandalwood, for example are obtained by steam distillation. This method separates the essential oil, which are volatile and fairly insoluble in water, leaving the compounds like tannins, bitter principles, and mucilaginous constituents, in the distillation assembly. Essential oils so obtained are re-distilled to get rid of the extraneous compounds. The volatile oils may be in the liquid phase or solid form (orris) or in a semi-solid form (rose) depending on the prevailing temperature. The essential oils are soluble in fats as well as in ethyl alcohol. They volatilize without leaving any residue behind.

2.11.1 Concretes

The raw materials, like bark, leaf, and herb may be extracted with a hydrocarbon solvent (like petroleum or hexane). This method of extraction is carried out when the volatile oil is not stable to higher temperatures that prevail during hydro-or steam distillation. Additionally, the solvent extraction method produces essential oil, free form artifacts and the odor corresponds to the natural specimen. In the case of some fragrant herbs, both the methods are employed for isolation of essential oils. The examples are oils of lavender and sage. Concrete consists of a mixture of wax and essential oil, the relative percentage of these components varies from herb to herb. Thus, for concrete derived from jasmine the relative proportion is 50:50, while for Ylang Ylang, the concrete is a liquid that comprises of 80% essential oil, while the rest is wax (20%). One outstanding advantage associated with concretes is that these are quite stable and more concentrated in terms of essential oil content.

2.11.2 Resinoids

The hydrocarbon solvent extract of resinous material is termed as a "resinoid". Typical examples are benzoin or Peru balsam. The alcohol soluble fraction of a resinoid is termed as 'Absolute'. Resinoids are useful as fixative to prolong the effectiveness of an essential oil.

2.11.3 Absolutes

Absolutes are obtained from concretes by extraction with pure alcohol, which removes most of the accompanying wax. The process may be repeated several times to bring the concentration level of wax to the minimum possible level. The absolute may be distilled under mild reduced pressure to recover alcohol from the absolute. To rid absolute of other unwanted impurities, it may be subjected to molecular distillation (using very high vacuum). The material thus obtained may still contain traces of ethyl alcohol (~2%) and therefore is not recommended for therapeutic use. The latest development in the extraction of essential oil is the employment of liquid carbon dioxide. The outstanding advantage is that the quality of the isolated essential oil is excellent and is completely free from the above described contaminants.

2.11.4 Effleurage

This is the method of extraction of essential oil from freshly cut flowers. It consists of a glass plate (20 cm × 20 cm) that is coated with a thin layer of odorless fat. A layer of aromatic flowers is laid on top of the fat layer and allowed to stand. The fatty layer that becomes saturated with the essential oil from the flowers (jasmine or tuberose) is called 'pamade'. The fat is then extracted from the pomade with ethyl alcohol. The evaporation of the solvent yields pure absolute or perfume.

The naturally occurring perfumes such as rose oil contain a large number of constituents (~300), some of which are still unidentified. The synthetic products corresponding to natural specimen of rose differ in odor and are not recommended for therapeutic use since they may cause allergies and distress to sensitive skin. The difference in the fragrance is that genuine rose oil contains many desirable trace ingredients that are lacking in the laboratory synthesized specimen.

2.11.5 Miscellaneous applications

Some of the problems (and their solutions) encountered by people.

Travel: Problems encountered during air journey include swelling in the ankles, feet, headache, and dry skin. The causative reason is the pressurized cabins in the airplane, leading to dehydration of the passenger followed by the above-mentioned ailments. To avoid this situation, one should refrain from consuming tea, coffee, and alcoholic beverages during a flight. In the event of already existing stomach problems, consumption of the above-mentioned beverages might cause bloating due to expansion of gas present in the stomach. It is recommended that the passenger, before the start of the air journey, consume a cup of peppermint tea, fortified with honey. Second, the passenger should carry a small handkerchief slightly wetted with water containing a few drops of lavender oil. The inhalation of the essential oil provides some relief to the passenger. To provide relief from swelling in the ankles and feet, a compress containing geranium oil can be applied to the affected area and then massaged in the upward direction. To avoid contracting pathogenic microorganisms while staying in a hotel, one should disinfect the toilet seat by wiping it with a tissue impregnated with a mixture of lavender and thyme oils.

Fever: In the event one contracts fever while traveling, it is strongly recommended that the patient takes complete rest for a few days. For bringing down the body temperature to normal level, sponging should be done with water containing a mixture of eucalyptus, lavender and thyme. Additionally, the patient should recuperate in fresh air and take plenty of fluids, including water and fruit juices. The recovery room may be sterilized by spraying with essential oil derived from thyme.

Dangerous insect bites: Mosquitoes, for example, spread killer diseases like malaria and dengue. The best option is to smear the exposed parts of the body with commercially available mosquito repellant preparations or keep lemongrass or citronella oil ready at hand and use them when sleeping in a mosquito infested area.

***Itch Producing plants*:** Stinging nettle plant in the event of accidental contact with the skin produces intense inching. The problem area should be washed with water, followed by a cold compress containing either of the following essential oils (chamomile, eucalyptus, lavender). The skin irritation stops within a short time.

***Workplace*:** The confined spaces of the modern-day offices are the breeding ground for various bacteria and viruses. This makes office workers susceptible to illnesses like cough and common cold, leading to a drop in the working efficiency of the organization. The various sources of health hazards include, for example, faulty humidifiers attached to air conditioners, poorly maintained photocopiers (that are reported to generate small quantities of ozone and nitrogen dioxide, an atmospheric pollutant), electromagnetic radiations emitted from the fluorescent screens of desktops and other electronic gadgetry. In the interest of the health of the office workers, it is recommended that citrus essential oil should be sprayed in the office from time to time. This is to clear up the stale air and create a calm and tranquil atmosphere. It is also advisable to have large leaved green plants in the office as they improve air quality. All these steps will increase the working efficiency of the organization.

***Facing stress*:** The inhalation of some essential oil is desirable, particularly when one is going for a job interview or appearing for an examination. The oils boost confidence, enhance memory, and enable the person to have a sharp focus on the problem on hand. Some essential oils that help dealing with stress by calming the nerves are basil, chamomile, lavender, and sandalwood.

Senior citizen problems: Elderly people experience a great deal of discomfort, particularly in their hands and feet, due to sluggish blood circulation. To overcome this condition, it is recommended to rub geranium oil (diluted in vegetable oil) on their body. This treatment makes the skin smoother, firmer, and brings a healthy glow on the face. It may be recalled that geranium oil also exerts an antidepressant action on the human system. Lying down with legs raised (such that they are located above the heart level) for half an hour, will provide relief for swollen ankles and feet. During summertime, the problem is solved as follows: a piece of ice is put in a polythene bag and rubbed on the backside of knees till these get cold and then it is applied over the center of the soles of feet. Subsequently the mixture of two essential oils (fennel and cypress oils diluted with vegetable oil) is massaged upward from feet to back and front of the knees.

Leg cramps: One method of getting relief for leg cramps is by using Chinese acupressure. The method consists of holding firm the big toe between the thumb and the index finger for some time. Providing warmth and oil-massage are alternative methods to alleviate the symptoms. Thus, before going to bed, the whole leg should be massaged (with the massage oil composition, *vide infra*) in an upward direction. The massaging of the feet should follow next. The massage oil for leg cramps is a mixture of hyssop, lavender, marjoram and rosemary, diluted with vegetable oil. A pair of socks should be worn to maintain warmth in the leg muscles.

***Insomnia*:** With the advent of new imaging technology like photon microscopy, scientists for the first time can conclude that the human brain gets rid of toxins during sleep. Accumulation of toxins in the brain are responsible for Alzheimer's disease (dementia) and other neurological conditions. Therefore, it is imperative to get good sleep for keeping our brain in good health. Uninterrupted sleep for eight hours is considered ideal, but with the advancement of age this period usually is shortened. That should not be a cause of worry. The problem of sleeplessness however needs immediate attention. The followings tips for a good sleep may be found useful:

- Discipline oneself in regulating the time for going to bed and waking up.
- Head and neck massage with the massage oil (*vide infra*) helps in inducing sleep.
- Consuming a cup of warm milk or chamomile tea before retiring to bed is found to be relaxing. Caution should, however, be exercised not to consume too much liquid in the evening, otherwise one may have to get up, couple of times, during night to empty one's bladder.
- Practitioners of aromatherapy recommend the use of following synergic blends for foot massage for alleviating the problem of sleeplessness (viz; chamomile, lavender, marjoram, and valerin). Foot contains seventy-two thousand nerve endings, which is why foot massage makes the entire body feel so wonderfully relaxed.
- Meditation in a well-ventilated room helps a person achieve a state of peace and tranquility. This helps restore energy and induce sound sleep thereafter.

Loss of memory: Forgetfulness is experienced at all stages of life. It arises due to lack of concentration and having too much on one's mind, or thinking of multiple things at a time. To help restore concentration, one is advised to use the following essential oils (taken in equal proportion- for inhalation as such or put in an atomizer and sprayed several times a day in the living room.) viz; basil, black pepper, cardamom, and ginger. Additionally, the affected persons should take orally a multivitamin capsule, that contains vitamins, minerals, and micronutrients, including selenium and zinc. People suffering from Alzheimer's disease suffer from dementia and the condition arises due to the deterioration of the neurons in the brain. Recent research findings suggest that aluminum ions are the culprit for the said disorder. These ions find their way into the human system either through aerosol sprays or drinking water. The departments of public water distribution system use (unwittingly though!) aluminum salt for purification of water. There is, therefore, an urgent need for an effective alternative method for water purification. Additionally, the use of aluminum cooking pots should be abandoned altogether. Aluminum that has already entered the human body can be reduced gradually by oral intake of vitamin C.

Stem cell transplantation is one good option for curing Alzheimer's disease. However, for less advanced cases — the use of essential oils (viz; basil, geranium, lavender, rosemary with vegetable oil as a diluent) as ingredients for a full body massage is recommended. This may be supplemented with the following additional essential oils, which serve as stimulant to the human system. (viz; cardamom, coriander, ginger, grape-fruit, lemon, nutmeg, and orange).

2.12 Essential Oils Derived from Medicinal Herbs Used for the Treatment of Following Disorders:

2.12.1 Skin

➢ Black Pepper

- Slack tissues
- Chilblains

➢ Chamomile

- Eczema
- Insect bites
- Inflamed skin
- Psoriasis
- Tooth and gum ache
- Wounds
- Allergies
- Hair care
- Blisters, boils
- Sores
- Dermatitis
- Dry skin
- Chilblains

➢ Citronella

- Excessive perspiration

➢ Eucalyptus

- Insect bites
- Spots on the skin
- Athletic foot
- Abscesses/boils, blisters

- Burns
- Dandruff

➢ Jasmine

- Dry/oily skin
- Inflamed skin and
- Old age wrinkles

➢ Lavender

- Eczema
- Oily skin
- Insect bites
- Inflamed skin
- Rashes
- Wrinkled skin
- Allergies
- Abscesses and blisters
- Bruises
- Cuts/Sores
- Dermatitis
- Dry and sensitive skin

➢ Lemon grass

- Excessive perspiration
- Oily skin
- Scabies
- Acne
- Athlete's foot
- Slack tissue

➢ Marigold

- Oily or greasy skin

- Inflamed skin
- Old age wrinkles

➢ Patchouli

- Eczema
- Scars or stretch marks
- Dandruff

➢ Peppermint

- Ringworm
- Scabies
- Tooth ache
- Acne
- Dermatitis

➢ Sandalwood

- Oily scalp
- Rashes
- Scars
- Old age wrinkles
- Chapped and cracked skin

➢ Spearmint

- Ring worm
- Acne
- Scabies
- Congested and dull skin
- Dermatitis
- Tooth ache

➢ Thyme

- Mouth and gum infections

- Scabies
- Abscesses, blisters
- Bruises
- Insect bites
- Lice

➢ Turpentine

- Lice
- Ringworm
- Wounds
- Abscesses and blisters,

2.12.2 Circulation, muscles, and joints

➢ Aniseed

- Rheumatism

➢ Basil

- Gout
- Rheumatism

➢ Black Pepper

- Aches and Pains
- Arthritis
- Poor muscle tone
- Muscular cramps and stiffness
- Poor circulation and low blood pressure
- Rheumatism
- Sprains and strains

➤ Borneol

- Poor circulation and low blood pressure

➤ Camphor

- Aches and pains
- Arthritis
- Rheumatism
- Sprain and strain

➤ Chamomile

- Aches and pains
- Arthritis
- Rheumatism
- Sprains and strains

➤ Cinnamon

- Poor circulation and low blood pressure

➤ Clove buds

- Rheumatism
- Sprains and strains

➤ Coriander

- Accumulation of toxins
- Aches and pain
- Arthritis
- Gout
- Muscular cramps and stiffness
- Low blood pressure and poor circulation

➢ Cumin

- Accumulation of toxins
- Poor circulation and low blood pressure

➢ Eucalyptus

- Aches and pains
- Arthritis
- Poor circulation, low blood pressure
- Rheumatism

➢ Garlic

- High blood pressure, hypertension

➢ Ginger

- Poor muscle tone
- Poor circulation, low blood pressure
- Rheumatism
- Sprains and strains.

➢ Jasmine

- Muscular cramp and stiffness
- Sprains and strains

➢ Juniper

- Accumulation of toxins
- Obesity

➢ Lavender

- Aches and Pains
- High blood pressure, hypertension
- Muscular cramps and stiffness

- Rheumatism
- Sprains and strains

➢ Lemon

- Cellulites
- High bold pressure, hypertension,
- Obesity
- Poor circulation and low blood pressure

➢ Lemon grass

- Aches and pains
- Poor circulation and low blood pressure

➢ Nutmeg

- Poor circulation and low blood pressure.

➢ Orange (bitter and sweet)

- Edema and water retention
- Obesity
- Palpitation

➢ Orange blossom

- Palpitation
- Poor circulation and low blood pressure.

➢ Parsley

- Accumulation of toxins
- Cellulites
- Rheumatism

- Peppermint

  - Aches and pains
  - Arthritis
  - Rheumatism
  - Sprains and strains

- Pine

  - Accumulation of toxins

- Rose

  - Low blood pressure, poor circulation

- Sage

  - Edema and water retention
  - High blood, hypertension
  - Poor circulation and low blood pressure
  - Rheumatism

- Spearmint

  - Aches and pains

- Sweet funnel

  - Accumulation of toxins
  - Obesity
  - Rheumatism

- Turmeric

  - Aches and pains
  - Arthritis
  - Rheumatism
  - Sprains and strains

➢ Thyme

- Cellulites
- Gout
- Muscular cramps and stiffness
- Poor circulation and low blood pressure
- Sprains and strains

➢ Vetiver

- Sprains and strains

➢ Violet

- Poor circulation and low blood pressure
- Rheumatism

➢ Yarrow

- High blood pressure, hypertension
- Rheumatism

➢ Ylang Ylang

- High blood pressure, hypertension
- Palpitation

2.12.3 Digestive system

➢ Anise

- Indigestion/flatulence

➢ Aniseed

- Cramps/gastric spasm
- Indigestion/flatulence

➢ Basil

- Indigestion/flatulence
- Nausea/vomiting

➢ Black Pepper

- Colic
- Constipation and sluggish digestion
- Cramps/gastric spasm
- Heartburn
- Indigestion and flatulence
- Loss of appetite
- Nausea/vomiting

➢ Caraway

- Colic
- Cramps/gastric spasm
- Indigestion/flatulence
- Loss of appetite

➢ Celery seeds

- Liver congestion

➢ Chamomile

- Colic
- Indigestion/flatulence
- Nausea/vomiting

➢ Cinnamon

- Constipation
- Cramps/gastric spasm
- Indigestion/flatulence

- Clove Buds
  - Indigestion/flatulence
  - Nausea/vomiting
- Cardamom
  - Colic
  - Cramps/gastric spasm
  - Griping pain
  - Indigestion/Flatulence
  - Loss of appetite
  - Nausea/vomiting
- Coriander
  - Colic
  - Cramps/gastric spasm
  - Indigestion/flatulence
  - Nausea/vomiting
- Cumin
  - Colic
  - Cramps/gastric spasm
  - Indigestion, flatulence
- Dill
  - Colic
  - Griping pain
- Ginger
  - Colic
  - Cramps/gastric spasm
  - Indigestion/flatulence
  - Loss of appetite
  - Nausea/vomiting

➤ Hops

- Indigestion/flatulence

➤ Lavender

- Colic
- Cramps/gastric spasm
- Indigestion/flatulence
- Nausea/vomiting

➤ Lemon

- Gastric spasm/cramps

➤ Lemongrass

- Indigestion/flatulence

➤ Nutmeg

- Constipation and sluggish constipation
- Indigestion/flatulence
- Nausea/vomiting

➤ Orange (Bitter sweet)

- Constipation and sluggish digestion
- Indigestion/flatulence

➤ Orange blossom

- Colic
- Cramps/gastric spasm
- Indigestion/flatulence

➤ Palmarosa

- Constipation and sluggish digestion

➢ Parsley

- Colic
- Griping pain
- Indigestion/flatulence

➢ Mint (Peppermint and spearmint)

- Colic
- Cramps/gastric spasm
- Indigestion/flatulence
- Nausea/vomiting

➢ Rose

- Liver congestion
- Nausea/vomiting

➢ Rosemary

- Colic
- Indigestion/flatulence
- Liver congestion

➢ Rose wood

- Nausea and vomiting

➢ Sage

- Colic

➢ Sandalwood

- Nausea/vomiting

➢ Sweet fennel

- Colic
- Constipation and sluggish digestion

- Griping pain
- Indigestion/flatulence
- Nausea/vomiting

➢ Thyme

- Indigestion/flatulence

➢ Turmeric

- Constipation and sluggish digestion
- Liver congestion

➢ Valerian

- Indigestion/flatulence

➢ Yarrow

- Constipation and sluggish digestion
- Cramps/gastric spasm
- Indigestion/flatulence.

2.12.4 Genito-urinary/Endocrine system

➢ Basil

- Lack of menstruation painful/difficult menstruation

➢ Black Pepper

- Frigidity

➢ Cedarwood

- Leucorrhea
- Itching

- Celery seeds

    - Cystitis
    - Lack of nursing

- Chamomile

    - Cystitis
    - Painful or difficult menstruation
    - Excessive menstruation

- Cinnamon leaf

    - Lack of menstruation
    - Frigidity
    - Labor pain and childbirth aid
    - Leucorrhea

- Dill

    - Lack of menstruation
    - Lack of nursing milk

- Eucalyptus: (Blues gum)

    - Cystitis
    - Leucorrhea

- Frankincense

    - Cystitis
    - Painful or difficult menstruation
    - Leucorrhea

- Germanium

    - Thrush/candida

➢ Hops

- Lack of menstruation
- Painful or difficulty menstruation
- Lack of nursing milk
- Sexual overactivity

➢ Jasmine

- Painful or difficult menstruation
- Frigidity
- Labor pain and childbirth aid
- Menopausal problems

➢ Juniper

- Itching

➢ Lavender

- Cystitis
- Painful or difficult menstruation
- Labor pain and childbirth aid
- Leucorrhea
- Premenstrual syndrome
- Itching

➢ Myrrh

- Leucorrhea
- Itching
- Thrush/candida

➢ Nutmeg

- Frigidity
- Labor pain or childbirth aid

- Orange blossom
  - Frigidity
  - Premenstrual syndrome
- Parsley
  - Lack of menstruation
  - Cystitis
  - Frigidity
  - Labor pain and childbirth aid
- Patchouli
  - Frigidity
- Pine
  - Cystitis
- Rose
  - Lack of menstruation
  - Painful or difficult menstruation
  - Menopausal problems
  - Excessive menstruation
- Rosemary
  - Painful or difficult menstruation
  - Leucorrhea
- Rosewood
  - Frigidity
- Sage
  - Lack of menstruation
  - Painful or difficult menstruation
  - Frigidity

- Labor pain and childbirth aid
- Leucorrhea

➢ Sandalwood

- Cystitis
- Frigidity
- Leucorrhea

➢ Sweet Fennel

- Lack of menstruation
- Lack of nursing milk
- Menopausal problems

➢ Thyme

- Cystitis

➢ Turpentine

- Cystitis
- Leucorrhea
- Urethritis

➢ Yarrow

- Lack of menstruation
- Cystitis
- Painful or difficult menstruation

➢ Ylang Ylang

- Frigidity

2.12.5 Immune system

➢ Angelica

- Cold/flu

- Aniseed
  - Cold/flu
- Basil
  - Cold/flu, fever
- Borneol
  - Cold/flu, fever
- Camphor
  - Cold/flu, fever
- Caraway
  - Cold/flu, fever
- Chamomile
  - Chicken pox, fever
- Cinnamon
  - Cold/flu, fever
- Citronella
  - Cold/flu, fever
- Clove bud
  - Cold/flu, fever
- Coriander
  - Cold/flu, fever
- Eucalyptus
  - Chicken pox
  - Cold/flu, Fever

➤ Frankincense

- Cold/flu

➤ Ginger

- Cold/flu, fever

➤ Juniper

- Cold/flu, fever

➤ Lavender

- Chicken pox, measles

➤ Lemon

- Fever

➤ Lemon grass

- Fever

➤ Lime

- Fever

➤ Mint (Peppermint and spearmint)

- Cold/flu, fever

➤ Myrtle

- Fever

➤ Orange (sweet and bitter)

- Cold/flu

➤ Pine (long leaf)

- Cold/flu

➤ Thyme

- Fever

➤ Turpentine

- Cold and flu

➤ Yarrow

- Cold/flu, fever

2.12.6 Nervous system

➤ Angelica

- Migraine
- Nervous exhaustion or fatigue/ debility
- Nervous tension and stress

➤ Asafoetida

- Nervous exhaustion or fatigue/debility
- Nervous tension and stress

➤ Balsam

- Nervous tension and stress

➤ Basil

- Anxiety
- Depression
- Insomnia
- Migraine
- Nervous exhaustion or fatigue/debility
- Nervous tension and stress

➢ Borneol

- Nervous exhaustion or fatigue/debility
- Nervous tension and stress
- Neuralgia/sciatica

➢ Cardamom

- Nervous exhaustion or fatigue/debility
- Nervous tension and stress

➢ Cedarwood

- Nervous tension and stress

➢ Celery seeds

- Neuralgia and sciatica

➢ Chamomile

- Headache
- Insomnia
- Migraine
- Nervous tension/stress
- Neuralgia/sciatica

➢ Cinnamon

- Nervous exhaustion or fatigue/ debility Nervous tension and stress

➢ Citronella

- Headache
- Migraine
- Nervous exhaustion or fatigue/ debility
- Neuroglia/ Sciatica

➢ Cumin

- Headache
- Nervous exhaustion or fatigue/debility

➢ Cypress

- Nervous tension and stress

➢ Eucalyptus

- Headache
- Nervous exhaustion or fatigue
- Neuroglia/sciatica

➢ Frankincense

- Anxiety

➢ Geranium

- Nervous tension and stress
- Neuralgia/sciatica

➢ Ginger

- Nervous exhaustion or fatigue/ debility

➢ Hops

- Headache
- Insomnia
- Nervous tension and stress
- Neuralgia/sciatica

➢ Jasmine

- Anxiety
- Nervous tension and stress

➢ Lavender

- Depression
- Headache
- Insomnia
- Migraine
- Nervous exhaustion or fatigue/ debility
- Nervous tension and stress
- Neuralgia, sciatica
- Shock
- Vertigo

➢ Lemon

- Insomnia
- Nervous tension and stress

➢ Lemon grass

- Headache
- Nervous exhaustion or fatigue/ debility
- Nervous tension and stress

➢ Mint (Peppermint and spearmint)

- Headache
- Migraine
- Nervous exhaustion or fatigue/ debility
- Nervous tension and stress
- Neuralgia/sciatica
- Vertigo

➢ Nutmeg

- Nervous exhaustion or fatigue/ debility
- Neuralgia/sciatica

➢ Orange: (bitter and sweet)

- Nervous tension and stress

➢ Orange blossom

- Insomnia
- Nervous tension and stress
- Shock

➢ Palmarosa

- Nervous exhaustion or fatigue/debility
- Nervous tension and stress

➢ Patchouli

- Nervous exhaustion or fatigue/ debility
- Nervous tension and stress

➢ Peppermint

- Headache

➢ Pine

- Nervous exhaustion or fatigue/debility
- Nervous and stress
- Neuralgia/sciatica

➢ Rose

- Headache
- Nervous tension and stress

➢ Rosemary

- Headache
- Nervous exhaustion and fatigue/ debility
- Nervous tension and stress
- Neuralgia/sciatica

➢ Rosewood

- Headache
- Nervous tension and stress

➢ Sage

- Depression
- Headache
- Migraine
- Nervous exhaustion and fatigue/ debility
- Neuralgia/sciatica

➢ Sandalwood

- Depression
- Insomnia
- Nervous tension and stress

➢ Thyme

- Headache
- Insomnia
- Nervous exhaustion or fatigue/debility
- Nervous tension and stress

➢ Turpentine

- Neuralgia/sciatica

➢ Valerian

- Insomnia
- Migraine
- Nervous tension and stress

➢ Vetiver

- Depression
- Insomnia
- Nervous exhaustion or fatigue/debility

➢ Violet

- Headache
- Insomnia
- Nervous exhaustion or fatigue/debility
- Nervous tension/stress
- Vertigo

➢ Yarrow

- Insomnia
- Migraine
- Nervous tension and stress

➢ Ylang Ylang

- Anxiety
- Depression
- Insomnia
- Nervous exhaustion or fatigue/debility
- Nervous tension and stress

Chapter 3

Native Indian Plants (171) and Their Biological Activities with Respect to Human Body

3.1 Introduction

The use of medicinal plants has a long tradition in India. The bedrock of this tradition is Ayurveda, the ancient Indian system of medicine. An important branch of Ayurveda pertains to plant-based medicines. Given the long standing and rich tradition of medicinal plant usage in India, plants native to India offer an excellent source of study for researchers that want to undertake the scientific study of plants used in alternative medicine systems. Accordingly, in this chapter we document 171 notable Indian medicinal plants. These plants are documented in tabular form. The table is to be read as follows:

- † The medicinal plants are numbered in Arabic numerals in going from left to right of each page of the table.
- ≠ Properties associated with each medicinal plant are represented in Roman numerals and are placed in a sequential order arranged from top to bottom of the page.
- Σ The botanical names of the plants referred to as well as the number assigned to them is provided in the tables as well as appendix (pages 186–189).

A total of 55 biological properties with respect to the 171 Indian medicinal plants are described in the color-coded tables provided in this chapter. The 55 biological properties have been defined in section 3.2. The master table provides a quick overview of the properties exhibited by these 171 medicinal plants.

3.2 Definition of the Biological Properties Used in the Tables

This section provides glossary of the 55 biological properties used in the tables and is divided into five subsections (a-e).

3.2a

Glossary

- (I)   Abortifacient: Promotes abortion.
- (II) Alterative: Corrects the disordered processes of nutrition.

(III)   Anodyne: Pain reliever.

(IV)   Anthelmintic: Kills intestinal worms.

(V)   Antiasthmatics: Relief from disorder of bronchial tubes.

(VI)   Antibilious: Corrects secretion of bile.

(VII)   Antidiabetic: Provides means of releasing correct amount of insulin.

(VIII)   Antidote: Counteracts or neutralizes the action of poison.

(IX)   Antidiarrhoeics and antidysenterics: Prevents diarrhea and dysentery.

(X)   Antiemetic: Prevents nausea and vomiting.

3.2b

Glossary

(XI)   Antilithics: Counteracts the formation of stone/calculus.

(XII)   Antiparasitic: Kills the organisms that obtain food or shelter from another host organism

(XIII)   Antiphlogistic: Counteracts inflammation.

(XIV)   Antiscorbutics: Prevents/cures scurvy.

(XV)   Antiseptic: Substances that destroy or inhibit the growth of microorganisms.

(XVI)   Antispasmodic: Counteracts spasmodic disorders.

(XVII)   Aphrodisiac: Promotes sexual desire.

(XVIII)   Aromatic: Fragrant/pleasing smell.

(XIX)   Astringent: Arrests secretion or bleeding.

(XX)   Bitters and Bitter tonics: Opposite of sweet.

3.2c

Glossary

(XXI)   Carminative: Relieve flatulence.
(XXII)   Demulcent: Soothing effect on skin.
(XXIII) Dentifrices: Powder/paste etc. for tooth cleaning.
(XXIV)   Deobstruents: Opening up of passages or pores of the body.
(XXV)   Depuratives: Purification and cleansing action on body.
(XXVI)   Diaphoretic: Induces copious perspiration.
(XXVII) Discutient: Disperses or absorbs a tumor.
(XXVIII) Diuretic: Increases the passage of urine.
(XXIX)   Emetic: Promotes vomiting.

3.2d

Glossary

(XXX)   Emmenagogue: Promotes menstruation.
(XXXI)   Emollients: Lessen skin irritation, swelling and pain.
(XXXII)   Expectorants: Promotes expulsion of phlegm.
(XXXIII) Eye drops and lotions: Provide relief from eye irritation.
(XXXIV)   Febrifuge: Reduces fever.
(XXXV)   Galactagogue: Promotes flow of milk.
(XXXVI)   Gargles: Washing one's throat with a liquid, keeping the liquid moving by breathing out through it.
(XXXVII) Hair tonic: Promotes healthy hair.
(XXXVIII) Liniments: Liquid for application on skin by massage.
(XXXIX)   Narcotic: Induces deep sleep.

3.2e

Glossary

- (XL) Pectoral: Cures chest diseases.
- (XLI) Purgative: Acts as a laxative.
- (XLII) Refrigerant: Relieves feverishness and produces a cooling effect.
- (XLIII) Rheumatism: Pain in the muscles, joints and certain tissues.
- (XLIV) Rubefacient: Mild counter-irritant causing redness of skin.
- (XLV) Sedative: Reduces excitement, irritation, and pain.
- (XLVI) Skin disease remedies: Something that lessens or cures skin problems.
- (XLVII) Stomachics: Strengthens and promotes stomach's action.
- (XLVIII) Styptics: Checks bleeding.
- (XLIX) Suppurative: Helps pus formation.
- (L) Tonic: Serves to invigorate.

3.2f

Glossary

- (LI) Urinary system diseases: Offers treatment for disorders of urinary organs.
- (LII) Uterine diseases: Useful for the treatment of diseases related to womb.
- (LIII) Venereal diseases: Remedy for diseases communicated by unprotected sexual intercourse.
- (LIV) Vesicant: Causes blistering.
- (LV) Vulneraries: Promotes healing of wounds.

3.3 Tabular Information of Medicinal Plants Showing Various Biological Properties

	Property	Plants
3.2a	Abortifacient	2, 14, 15, 16, 19, 27, 38, 40, 57, 68, 81, 95, 106, 111, 123, 131, 135, 138, 142, 143.
	Alterative	6–9, 11, 13, 15, 17–19, 23, 29, 32, 38, 40, 43–45, 47, 54, 67, 69, 78–79, 84, 89, 92–93, 95, 102–103, 109, 122, 127, 135–136, 140, 141–142, 145–146, 156–158, 163.
	Anodyne	37, 39, 65, 72, 89, 106, 114, 116, 131, 146.
	Anthelmintic	3–4, 6–9, 12–14, 16, 17, 20–22, 27, 29, 34–36, 38, 40, 41, 43, 48, 51, 57, 59, 63–64, 66, 69–70, 72, 81, 86, 88, 106–107, 109, 112–115, 121, 127–128, 130, 134, 137, 139, 141–144, 152–153, 156, 159, 166–167, 169.
	Antiasthmatic	2–6, 22, 24, 27, 31, 35, 38–39, 57, 65, 67, 72, 76, 79, 86, 88–89, 95, 105–106, 122.
	Antibilious	5, 20, 26, 30, 32, 41, 51, 61, 70–71, 74, 78, 86, 104, 113, 116–117, 128–129, 131, 138, 145, 152, 154–156, 161, 168–171.
	Antidiabetic	8, 11, 32, 34, 42–44, 59, 70, 73–74, 82–83, 99, 109, 127, 148, 151, 161.
	Antidote	3, 7, 9, 11–12, 19, 22, 24–25, 32, 71, 89, 109, 123, 139, 152.
	Antidiarrhoeics and Antidysenterics	1–5, 9, 14, 17, 19, 26–28, 30, 33–34, 38, 44–47, 49–51, 58–59, 61, 63, 66, 68, 70–71, 78–79, 81, 84–86, 92, 95, 97, 99, 103, 105, 108-110, 114, 117, 119, 121, 123–124, 126–127, 129–131, 135–136, 138–140, 143–147, 149–152, 156–157, 170–171.

(Continued)

(*Continued*)

	Property	Plants
	Antiemetic	8, 22, 25, 30, 36, 39, 46, 49, 70, 73, 78, 84, 108, 110, 116, 123, 148, 152, 156, 170.
3.2b	Antilithics	3, 16–17, 22, 25, 33–34, 47, 54, 62, 66, 70, 74, 81, 84, 86, 89, 94, 101, 103, 106, 118, 130, 132, 136, 154, 160, 163–164.
	Antiparasitic	3, 4, 12, 14, 25, 57, 82, 101, 126, 139, 141, 165, 167, 169.
	Antiphlogistic	4, 6, 18, 20–22, 31–32, 34–35, 45, 51, 59, 65, 67, 74, 82, 84, 95, 97, 102–103, 106, 111, 121, 124, 126, 129, 134, 137, 144, 147, 151, 161–162, 169.
	Antiscorbutics	5, 11, 35, 40, 71, 74, 95, 99, 104, 106, 109, 117, 127, 131–132, 137, 145, 161.
	Antiseptics	4, 6, 20, 25, 33, 46, 49, 51, 114, 120, 121–122, 126, 138, 141–142, 152.
	Antispasmodic	3–4, 15, 22–25, 31, 38–39, 53, 56, 61, 65, 67, 72, 86, 93, 101, 106, 118, 131, 135, 141–142, 156, 159, 164, 166.
	Aphrodisiac	3, 6–8, 10, 15–16, 18, 23, 27, 34, 39, 43, 65, 77, 81, 93, 95, 106–107, 109, 118–119, 121, 128, 131, 136, 141, 144–145, 148, 158, 160–161, 169.
	Aromatic	3–5, 8, 46, 53–54, 59–60, 65, 68, 71, 105, 114, 120–121, 123, 137, 141, 161, 166, 170.
	Astringent	1–2, 5, 7, 9, 14–16, 23, 28–29, 33–35, 37, 39, 44, 46–47, 49–50, 51–53, 58, 62–63, 65–66, 70–71, 73–76, 83, 86, 90, 93–94, 96–99, 102, 107, 109–110, 113, 116, 121, 124, 126–127, 129–131, 135, 138–140, 141, 144–145, 147–148, 150–157, 160–161, 164, 167–168, 171.

(*Continued*)

(Continued)

Property	Plants
Bitter and Bitter Tonics	3–4, 13, 18, 25, 36, 51, 70, 78, 86, 101–102, 113, 122, 137–138, 148, 171.
3.2c Carminative	3, 6, 8, 13, 15–16, 19, 25, 40, 46, 50–51, 53, 56, 58–61, 65, 68, 71–72, 74, 76, 86, 88, 100, 102, 112, 114, 116, 121–123, 126, 132, 135, 137, 151–152, 161, 166–167, 170.
Demulcent	5, 10, 17, 23, 27, 44, 48, 50–52, 54, 71, 79–81, 83–85, 96–97, 102, 109–110, 114–116, 118–119, 124, 136, 143, 145, 149, 160–161, 168.
Dentifrices	16, 32, 49, 156.
Deobstruents	19, 21–22, 37, 67, 88–89, 91, 101, 105, 123, 128, 135, 137–138, 163, 166, 169.
Depuratives	2, 29–30, 34, 43, 45, 47, 60, 70, 88, 95, 141, 171.
Diaphoretic	6, 30–31, 36, 38, 50, 55, 57, 63, 72–73, 76, 84, 88, 100, 113–114, 116, 120, 128–129, 131, 135, 143–144, 146, 152–153, 168.
Discutient	6, 20, 25, 35, 39, 57–58, 65, 67, 74, 92, 100–101, 106, 111, 116, 121, 123, 135, 138–139, 151, 161, 167.
Diuretic	2–3, 6, 10–11, 15, 17–18, 23, 25–26, 34, 38–39, 43–45, 47, 53, 57–58, 60–63, 66, 68, 70, 72, 77, 84–85, 88, 91, 95–96, 99, 101–102, 106–107, 109–110, 112–114, 118–119, 122–124, 127, 131–132, 137–138, 141, 143, 146–147, 161, 163–164, 166–170.

(Continued)

(*Continued*)

	Property	Plants
	Emetic	2–3, 17, 25, 28, 37–38, 44, 48, 55, 57, 67, 97, 100, 103–106, 131, 134, 136, 139, 163, 168.
3.2d	Emmenagogues	3, 6–7, 9, 19–20, 22, 25, 27, 31, 35, 40, 50, 58, 63, 66, 68, 72, 80–81, 88, 91, 94, 100–102, 105, 107, 112, 118, 122–123, 130–134, 137, 143, 161–164, 166, 170.
	Emollient	2, 11, 25, 35, 42, 44, 51, 53, 70, 85, 108, 114, 122, 124, 127, 129, 134, 138, 143, 153, 161–162, 168.
	Expectorants	2–4, 6, 17, 22–23, 25, 27–28, 38, 43, 50, 60, 72, 76, 79, 81, 86, 88, 97, 102, 110, 113–114, 131, 135, 139, 141, 146, 163.
	Eyedrops and Lotions	2, 5, 16–17, 31, 36–37, 48, 53, 59, 70, 84, 88, 90, 102, 111, 119, 121, 129–130, 133–134, 136, 145, 149–50, 152–153, 155–156, 170–171.
	Febrifuge	4–6, 13, 19, 25–28, 35, 38, 40, 42–45, 54–55, 60, 66, 70, 78, 81, 85–86, 92, 100, 102, 105–106, 108, 113–114, 119, 125–126, 130, 132–134, 145, 147–148, 152, 154, 168–170.
	Galactagogue	9–10, 23, 36, 58, 63, 77, 80–81, 92, 95, 104, 112, 134, 143, 161, 171.
	Gargles	32, 49–50, 52–53, 69–70, 74–76, 86, 88, 92–94, 97, 99, 106, 109, 118, 121, 123, 130, 146–147, 150–151, 156, 160.
	Hair Tonics	14, 25, 39, 49, 57, 65, 67, 94, 123, 130, 141, 143, 153, 155, 161, 166.

(*Continued*)

(Continued)

Property	Plants
Liniments	3, 6, 16, 25–26, 35, 37, 46, 54, 57, 65, 79, 81–82, 88–89, 92, 96, 102, 104, 106, 126, 152, 155.
Narcotics	12, 17, 39, 56, 65, 117, 170.
3.2e Pectoral	1–4, 6, 8, 17, 22, 26–28, 32, 38, 52, 56, 60, 67, 69, 76, 79, 81, 89–90, 95–96, 99, 106, 110, 114–115, 119, 121–126, 131, 134, 142–143, 154–155, 160–161, 163–164, 166, 169–171.
Purgative	2–5, 7, 9, 14–17, 25–26, 31–32, 37–38, 41–45, 48–52, 55, 57, 64, 67, 72, 75, 78–81, 88, 92–93, 95–99, 101–102, 104–106, 108–109, 113, 115, 119, 122–125, 128, 131–132, 134, 139, 143, 145–146, 150, 152, 155–156, 160, 163–164, 166, 168, 171.
Refrigerant	23, 25, 32, 49, 53, 61, 68, 70, 110, 115, 117, 119, 130, 152–153, 160, 162.
Rheumatism	2–4, 6–8, 11, 16, 20, 23–24, 26, 35, 37–38, 41, 45–46, 50, 53–54, 56, 61, 64–65, 67, 73–74, 82, 84, 92, 94–97, 100, 102, 104–107, 113, 116, 122, 124–126, 131, 134, 137, 142, 144–145, 147–148, 152, 158, 161, 163–164, 170–171.
Rubefacient	6, 11, 34, 49, 54–55, 57, 61, 92, 95, 106, 121–123, 125, 142, 164, 170.
Sedative	2, 24, 39, 46–47, 54, 56, 65, 72, 81, 100, 131–132, 135, 138, 140–142, 145–146, 153, 166.

(Continued)

(*Continued*)

Property	Plants
Skin disease (Remedies)	1–2, 4, 6, 8, 10–12, 17, 19–20, 25, 27, 29–34, 37–38, 40–43, 45, 47, 49–51, 59, 64–65, 67, 69–70, 72, 75, 78, 81–85, 87, 89–90, 92–99, 101, 104, 108, 110–111, 113–115, 120, 123, 125–129, 133–136, 138, 141–143, 146, 148, 152–153, 157–158, 162–165, 167, 169.
Stomachics	2, 5–6, 8, 13, 17, 20–22, 25, 27, 30–31, 38–40, 47, 50–51, 54, 56, 58, 60–61, 63, 65, 69–71, 74, 76, 78, 80, 86, 88, 102, 104, 114, 116–117, 120, 122–123, 126, 128, 130, 134, 141, 145, 148, 150–151, 156, 158, 166–167, 170.
Styptics	1, 22, 32–33, 44, 49, 59, 62, 67, 92, 96, 99, 109–110, 127, 130, 136, 164, 171.
Suppurative	14, 17–18, 75, 81, 91, 96, 106, 115, 117, 133–134, 145, 149.
Tonics	1, 3, 6–8, 15–23, 25–30, 32, 34, 36, 38–39, 41, 44–47, 50–56, 58–61, 66–72, 74, 76, 78, 80–82, 84, 86, 88, 92–93, 95–97, 99–100, 102–107, 109–110, 112–114, 116, 119, 121–123, 126, 128–132, 135–148, 152, 154–161, 163, 166–167, 169–170.
3.2f Urinary system diseases	8, 16, 22, 32, 37, 51, 54, 59, 67, 79–80, 86, 89–90, 94, 130, 135–136, 152–153, 156, 158.
Uterine diseases	1–3, 10–11, 20, 22, 28, 35, 46, 50, 52, 88, 94, 109, 118, 130, 135, 140, 150, 162, 169.

(*Continued*)

(*Continued*)

Property	Plants
Venereal diseases	1–2, 4, 10, 17, 29, 31–32, 37–38, 43–45, 48–49, 58–60, 62, 67, 70, 74–75, 79–80, 82, 84, 93, 95–96, 99, 102, 106, 109, 115, 118–120, 122–126, 131–132, 136, 138, 142, 144–146, 149, 157–158, 160, 165, 171.
Vesicant	6, 11, 32, 40, 54, 107, 137, 142.
Vulneraries	35–39, 63, 81, 88–89, 94, 96, 105–106, 133–134, 145, 154, 165.

3.4 Tables

A compendium of Indian medicinal plants and their uses are provided in a color coded tabular form from next page onwards.

Tables from pages 123–139 list the 171 plants with respect to the 10 properties described in section 3.2a.

Tables from pages 140–156 list the 171 plants with respect to the 10 properties described in section 3.2b.

Tables from pages 157–173 list the 171 plants with respect to the 9 properties described in section 3.2c.

Tables from pages 174–190 list the 171 plants with respect to the 10 properties described in section 3.2d.

Tables from pages 191–207 list the 171 plants with respect to the 11 properties described in section 3.2e.

Tables from pages 208–224 list the 171 plants with respect to the 5 properties described in section 3.2f.

Plants/Name/Uses	1 Catechu Tree	2 Rough-chaff/Latjira	3 sweet flag/Vacha	4 Malabar nut/Vasaka	5 Bael	6 Garlic/Lahsun	7 Aloe/Ghee Kunvar	8 Siamese Ginger/Kulanjan	9 Devil's Tree/Chatian	10 Cholai-bhajee
I Abortifacient		■ (green)								
II Alterative						■ (brown)	■ (brown)	■ (brown)	■ (brown)	
III Anodyne										
IV Anthelmintic			■ (orange)	■ (orange)		■ (orange)	■ (orange)	■ (orange)	■ (orange)	
V Antiasthmatics		■ (blue)	■ (blue)	■ (blue)	■ (blue)	■ (blue)				
VI Antibilious					■ (yellow)					
VII Antidiabetic										
VIII Antidote			■ (green)				■ (green)	■ (navy)	■ (green)	
IX Antidiarrhoeics and Antidysenteries	■ (grey)				■ (grey)				■ (grey)	
X Antiemetic								■ (purple)		

	11 Cashew nut/ Kaju	12 Fish berry/ Kakamari	13 The creat/ Kirata	14 Custard apple/ Sharifa	15 Celery/ Ajmoda	16 Areca nut/ Sufari	17 Prickly Poppy/ Bramha dana	18 Elephant Creeper/ Samudra-palaka	19 Indian Birthwort/ Isharmul	20 Wormwood/ Afsanthin
I Abortifacient				green	green	green			green	
II Alterative	brown		brown		brown		brown	brown	brown	
III Anodyne										
IV Anthelmintic		orange	orange	orange		orange	orange			orange
V Antiasthmatics										
VI Antibilious										yellow
VII Antidiabetic	dark blue									
VIII Antidote	green	green							green	
IX Antidiarrhoeics and Antidysenterics				grey			grey		grey	
X Antiemetic										

		21	22	23	24	25	26	27	28	29	30
	Plants/Name/ Uses	Santonica, Wormseed/ Kirmala	Indian Wormwood/ Nagadamani	Shatavari	Indiana Belladonna/ Angurshera	Nim Tree/ Nim	Thyme-leaved Gratiola/ Bamb.	Indian bamboo/ Bans	Indian Oak/ Sumudraphal	Variegated Bauhinia/ Kachanara	Barberry/ Rasaut
I	Abortifacient							green			
II	Alterative			brown						brown	
III	Anodyne										
IV	Anthelmintic	orange	orange					orange		orange	
V	Antiasthmatics		blue		blue			blue			
VI	Antibilious						yellow				yellow
VII	Antidiabetic										
VIII	Antidote		green		green	green					
IX	Antidiarrhoeics and Antidysenterics						tan	tan	tan		tan
X	Antiemetic		purple			purple					purple

Plants/Name/Uses		31 Pigweed/ Punarnava	32 Toddy Palm/Tad	33 Life plant/ Zakhme-hayat	34 Flame of the forest/ Palas	35 Fever nut/ Katkaranja	36 Calendula/ Zergul	37 Dilo oil Tree/ Champa	38 Swallow wort/ Madar	39 Indian Hemp/ Bhang, Charas	40 Papaya/ Papita
Abortifacient	I								green		green
Alterative	II		brown						brown		brown
Anodyne	III							peach		peach	
Anthelmintic	IV				orange	orange	orange		orange		orange
Antiasthmatics	V	blue				blue			blue	blue	
Antibilious	VI		yellow								
Antidiabetic	VII		dark purple		dark purple						
Antidote	VIII		olive green								
Antidiarrhoeics and Antidysenterics	IX			khaki	khaki				khaki		
Antiemetic	X						purple			purple	

	Plants/Name/ Uses	41 East Indian senna/ Senna	42 Purging cassia/ Amaltas	43 Senna Sophera/ Kasunda	44 White silk cotton Tree/ Sweta Salmali	45 Indian Pennywort/ Brahmi	46 Cinnamon/ Dalchini	47 Pareira brava / Path	48 Butterfly pea/ Aparajita	49 Coconut Palm/ Nariyal	50 Guggul
I	Abortifacient										
II	Alterative			■	■	■		■			
III	Anodyne										
IV	Anthelmintic	■		■					■		
V	Antiasthmatics										
VI	Antibilious	■									
VII	Antidiabetic		■	■	■						
VIII	Antidote										
IX	Antidiarrhoeies and Antidysenterics				■	■	■			■	■
X	Antiemetic						■			■	

Plants/Name/Uses		51 Jute Plant/ Kost	52 Indian Cherry/ Lasora	53 Coriander/ Dhania	54 Garlic pear/ Varuna	55 Poison bulb/ Kanwal	56 Saffron crocus/ Kesar	57 Purging croton/ Jamalgota	58 Cumin seed/ Jeera	59 Turmeric/ Haldi	60 Zedoary/ Gandhmul
Abortifacient	I							green			
Alterative	II				brown						
Anodyne	III										
Anthelmintic	IV	orange						orange		orange	
Antiasthmatics	V							blue			
Antibilious	VI	yellow									
Antidiabetic	VII									dark purple	
Antidote	VIII										
Antidiarrhoeics and Antidysenterics	IX	olive							olive	olive	
Antiemetic	X										

		61	62	63	64	65	66	67	68	69	70
	Plants/Name/ Uses	Lemon grass/ Bhustrina	Dhubgrass/ Dhub	Nut grass/ Nagar motha	Indian Rhubarb/ Nasatara	Indian thornapple/ Dhatura	Horse Gram/ Kulthi	Trailing Eclipta/ Bhringaraja	Lesser cardamom/ Choti- elachi	Embelia	Indian gooseberry/ Amla
I	Abortifacient								green		
II	Alterative							brown		brown	
III	Anodyne					peach					
IV	Anthelmintic			orange	orange		orange			orange	orange
V	Antiasthmatics					blue	blue	blue			
VI	Antibilious	yellow									yellow
VII	Antidiabetic										navy
VIII	Antidote										
IX	Antidiarrhoeics and Antidysenterics	grey		grey			grey		grey		grey
X	Antiemetic										purple

Plants /Name/ Uses		71 Wood apple/ Kaith	72 Asafoetida/ Hing	73 Indian Banyan/ Bargat	74 Indian fig tree/ Gular	75 Peepal tree/ Pipal	76 Indian Coffee Plum/ Talispatri	77 Fennel/ Barisauf	78 Chirata, Indian Gentian/ Bhuchiretta	79 Liquorice/ Mulahatti	80 Kashmir Tree / Gumbhar
Abortifacient	I										
Alterative	II								brown	brown	
Anodyne	III		tan								
Anthelmintic	IV		orange								
Antiasthmatics	V		blue				blue			blue	
Antibilious	VI	yellow			yellow				yellow		
Antidiabetic	VII			dark	dark						
Antidote	VIII	green									
Antidiarrhoeics and Antidysenterics	IX	grey							grey	grey	
Antiemetic	X			purple					purple		

Plants/Name/ Uses		81 Levant Cotton/ Kapaas	82 Chaulmugra seeds	83 East Indian screw Tree/ Mamorphali	84 Indian Sarsaparilla/ Anatmul	85 Lady Finger / Bhindi	86 Kurchi	87 Chaulmoogra	88 Hyssop	89 Indian indigo/ nil	90 Common Jasmine/ Chameli
Abortifacient	I	green									
Alterative	II				brown					brown	
Anodyne	III									tan	
Anthelmintic	IV	orange					orange		orange		
Antiasthmatics	V						blue		blue	blue	
Antibillious	VI						yellow				
Antidiabetic	VII		navy	navy							
Antidote	VIII									olive	
Antidiarrhoeics and Antidysenterics	IX	tan			tan	tan	tan				
Antiemetic	X				purple						

Plants/Name/Uses		91 Arabian Jasmine/Moghra	92 Purgingnut/Jamal-gota	93 Walnut/Akhrot	94 Henna plant/Mahendi	95 Garden Cress/Chandra-sur	96 Flax/Linseed/Alsi	97 Mohwa Tree/Mahua	98 Indian Kamila/Kamila	99 Mango Tree/Aam	100 Wild Chamomile/Babunah
Abortifacient	I					■					
Alterative	II		■	■		■					
Anodyne	III										
Anthelmintic	IV										
Antiasthmatics	V					■					
Antibilious	VI										
Antidiabetic	VII									■	
Antidote	VIII										
Antidiarrhoeics & Antidysenterics	IX		■			■		■		■	
Antiemetic	X										

Plants/Name/Uses		101 Indian Lilac	102 Yellow champa/Champake	103 Sensitive plant/Laifavanti	104 Bitter Gourd/Karela	105 Indian Mulberry/Shatoot	106 Horse raddish Tree/Sajana	107 Cowhage/Kavatch	108 Curry-leaf Tree/Katnimb	109 Banana/Kela	110 Sacred Lotus/Kamal
Abortifacient	I						green				
Alterative	II		brown	brown						brown	
Anodyne	III						peach				
Anthelmintic	IV						orange	orange		orange	
Antiasthmatics	V					blue	blue				
Antibilious	VI				yellow						
Antidiabetic	VII									navy	
Antidote	VIII									green	
Antidiarrhoeics and Antidysenterics	IX			tan		tan			tan	tan	tan
Antiemetic	X								purple		purple

Plants/Name/ Uses		111 Oleander/ Kanaer	112 Black Cumin/ Kalajira	113 Coral Jasmine/ Harsinghar	114 Holy Basil/ Tulsi	115 Prickly pear/ Chappal	116 Indian Trumpet flower/Jagdala	117 Yellow oxalis/ Chukatripatti	118 Bada gokhru	119 Date Palm/ Khajur	120 Longleaf Indian pine/ Chir
Abortifacient	I	(green)									
Alterative	II										
Anodyne	III				(peach)		(peach)				
Anthelmintic	IV		(orange)		(orange)	(orange)					
Antiasthmatics	V										
Antibilious	VI			(yellow)			(yellow)	(yellow)			
Antidiabetic	VII										
Antidote	VIII										
Antidiarrhoeics and Antidysenterics	IX				(gray)			(gray)		(gray)	
Antiemetic	X						(purple)				

		121	122	123	124	125	126	127	128	129	130
	Plants/Name/ Uses	Betel Leaf vine/ Pan	Long pepper/ Pippli	Black pepper/ kali mirch	Ispaghula/ Isabgul	Pagoda Tree/ Temple Tree/ Champa	Indian beech/ Karanj	Purslane/ Kurfa	Babchi seeds/ Bakuchi	Red Sandal wood/ Rakta-chandan	Pomegranate/ Anar
I	Abortifacient			green							
II	Alterative		brown					brown			
III	Anodyne										
IV	Anthelmintic	orange						orange	orange		orange
V	Antiasthmatics		blue								
VI	Antibilious								yellow	yellow	
VII	Antidiabetic							navy			
VIII	Antidote			green							
IX	Antidiarrhoeics and Antidysenterics			grey	grey		grey	grey		grey	grey
X	Antiemetic			purple							

Plants/Name/ Uses		131 Emetic nut/ Mainphala	132 Radish/ Muli	133 Indian Snakeroot/ Chota chand	134 Castor oil plant/ Erand	135 Indian madder/ Manjith	136 Silk Cotton Tree/ Shemul	137 Tooth brush tree/Pilu	138 Sandal Wood Tree/Safed Chandan	139 Three-leaf soapberry/ Ritha	140 Ashoka tree/ Ashoka
Abortifacient	I	green				green			green		
Alterative	II					brown	brown				brown
Anodyne	III	light orange									
Anthelmintic	IV				orange			orange		orange	
Antiasthmatics	V										
Antibilious	VI	yellow							yellow		
Antidiabetic	VII										
Antidote	VIII									green	
Antidiarrhoeics and Antidysenterics	IX	grey				grey	grey		grey		grey
Antiemetic	X										

Plants/Name/Uses	141 Costus root/Kuth	142 Marking nut tree	143 Sesame/Til	144 Shala tree/Sala	145 Country Mallow/Bala	146 Black nightshade/Makoy	147 Indian Red Wood Tree	148 Poison nut tree/Kuchala	149 Clearingnut tree/Nirmali	150 Lodh Tree/Lodh
I Abortifacient		green	green							
II Alterative	brown	brown			brown	brown				
III Anodyne						peach				
IV Anthelmintic	orange	orange	orange	orange						
V Antiasthmatics										
VI Antibilious					yellow					
VII Antidiabetic								navy		
VIII Antidote										
IX Antidiarrhoeics and Antidysenterics			olive	olive	olive	olive	olive	olive	olive	olive
X Antiemetic								purple		

Plants/Name/Uses		151 Jambolan/Jamun	152 Tamarind/Imli	153 Teak tree/Sagvan	154 Arjun Tree/Arjuna	155 Beleric myrobalan/Bahera	156 Chebulic myrobalan/Hirda	157 Tulip tree/Paras	158 Heart-leaved Moonseed/Amruta	159 Bishop's weed/Ajowan	160 Small caltrops/Chota gokhru
Abortifacient	I										
Alterative	II						■	■	■		
Anodyne	III										
Anthelmintic	IV		■	■			■			■	
Antiasthmatics	V										
Antibilious	VI		■		■	■	■				
Antidiabetic	VII	■									
Antidote	VIII		■		■						
Antidiarrhoeics and Antidysenterics	IX	■	■				■	■			
Antiemetic	X		■				■				

Plants/Name/Uses		161 Fenugreek/ Methi	162 Wheat/ Gehu	163 Indian squill/ Jangle piyaz	164 Stinging nettle/ Bichu	165 Indian copal tree/ Dhup	166 Indian Valerian/ Jata-mansi	167 Wild cumin/ Kali Zeera	168 Violet/ Banaf shah	169 Winter cherry/ Asvagandha	170 Ginger/ Sonth	171 Indian Date/ Ber
Abortifacient	I											
Alterative	II			brown								
Anodyne	III											
Anthelmintic	IV						orange	orange		orange		
Antiasthmatics	V											
Antibilious	VI	yellow							yellow	yellow	yellow	yellow
Antidiabetic	VII	dark										
Antidote	VIII											
Antidiarrhoeics and Antidysenterics	IX										olive	olive
Antiemetic	X										purple	

Plants/Name/Uses		Catechu Tree	Rough-chaff/Latjira	Sweet flag/Vacha	Malabar nut/Vasaka	Bael	Garlic/Lahson	Aloe (Ghee Kunvar	Siamese Ginger Kulanjan	Devil's Tree/Chatian	Cholai-bhajee
		1	2	3	4	5	6	7	8	9	10
XI	*Antilithics*			■ (green)							
XII	*Antiparasitic*			■ (brown)	■ (brown)						
XIII	*Antiphlogistics*				■ (peach)		■ (peach)				
XIV	*Antiscorbutics*					■ (orange)					
XV	*Antiseptic*				■ (blue)		■ (blue)				
XVI	*Antispasmodic*			■ (yellow)	■ (yellow)						
XVII	*Aphrodisiac*			■ (dark blue)			■ (dark blue)	■ (dark blue)	■ (dark blue)		■ (dark blue)
XVIII	*Aromatic*			■ (green)	■ (green)	■ (green)			■ (green)		
XIX	*Astringent*	■ (tan)	■ (tan)			■ (tan)		■ (tan)		■ (tan)	
XX	*Bitters and Bitter tonics*			■ (purple)	■ (purple)						

		11	12	13	14	15	16	17	18	19	20
Plants/Name/ Uses		Cashew nut/ Kaju	Fish berry/ Kakamari	The creat/ Kirata	Custard apple/ Sharifa	Celery/ Ajmoda	Areca nut/ Sufari	Prickly Poppy/ Bramha dana	Elephant Creeper/ Samudra-palaka	Indian Birthwort/ Isharmul	Wormwood/ Afsanthin
XI	*Antilithics*						■ (green)	■ (green)			
XII	*Antiparasitic*		■ (brown)		■ (brown)						
XIII	*Antiphlogistics*								■ (peach)		■ (peach)
XIV	*Antiscorbutics*	■ (orange)									
XV	*Antiseptic*										■ (cyan)
XVI	*Antispasmodic*					■ (yellow)					
XVII	*Aphrodisiac*					■ (navy)	■ (navy)		■ (navy)		
XVIII	*Aromatic*										
XIX	*Astringent*				■ (grey)	■ (grey)	■ (grey)				
XX	*Bitters and Bitter tonics*			■ (purple)					■ (purple)		

Plants /Name/ Uses	21 Santonica, Wormseed/ Kirmala	22 Indian Wormwood/ Nagadamani	23 Shatavari	24 Indiana Belladonna/ Agnurshera	25 Nim Tree/ Nim	26 Thyme-leaved Gratiola/ Bamb	27 Indian Bamboo/ Bans	28 Indian Oak/ Sumudraphal	29 Variegated Bauhinia/ Kachanara	30 Barberry/ Rasaut
XI *Antilithics*		green			green					
XII *Antiparasitic*					brown					
XIII *Antiphlogistics*	orange	orange								
XIV *Antiscorbutics*										
XV *Antiseptic*					blue					
XVI *Antispasmodic*		yellow	yellow	yellow	yellow					
XVII *Aphrodisiac*			dark				dark			
XVIII *Aromatic*										
XIX *Astringent*			olive					olive	olive	
XX *Bitters and Bitter tonics*					purple					

Plants/Name/Uses	31 Pigweed/ Punarnava	32 Toddy Palm/ Tad	33 Life plant/ Zakhme-hayat	34 Flame of the forest/ Palas	35 Fever nut/ Katkararnja	36 Calendula/ Zergut	37 Dilo oil Tree/ Champa	38 Swallow wart/ Madar	39 Indian Hemp/ Bhang, Charas	40 Papaya/ Papita
XI Antilithics			green	green						
XII Antiparasitic										
XIII Antiphlogistics	orange			orange	orange					
XIV Antiscorbutics					orange					orange
XV Antiseptic			cyan							
XVI Antispasmodic	yellow							yellow	yellow	
XVII Aphrodisiac				dark					dark	
XVIII Aromatic										
XIX Astringent			tan	tan	tan		tan		tan	
XX Bitters and Bitter tonics						purple				

Plants/Name/Uses		41 East Indian senna/ Senna	42 Purgining cassia/ Amaltas	43 Senna Sophera/ Kasunda	44 White silk cotton Tree/ Sveta Salmali	45 Indian Pennywort/ Brahmi	46 Cinnamon/ Dalchini	47 Pareira brava/ Path	48 Butterfly pea/ Aparajita	49 Coconut Palm/ Nariyal	50 Guggul
XI	Antilithics							■ (green)			
XII	Antiparasitic										
XIII	Antiphlogistics					■ (orange)					
XIV	Antiscorbutics										
XV	Antiseptic						■ (blue)			■ (blue)	
XVI	Antispasmodic										
XVII	Aphrodisiac			■ (dark)							
XVIII	Aromatic						■ (green)				
XIX	Astringent				■ (gray)		■ (gray)	■ (gray)		■ (gray)	■ (gray)
XX	Bitters and Bitter tonics										

Plants/Name/Uses		51 Jute Plant/ Kost	52 Indian Cherry/ Lasora	53 Coriander/ Dhania	54 Garlic pear/ Varuna	55 Poison bulb/ Kanwal	56 Saffron crocus/ Kesar	57 Purging croton/ Jamalgota	58 Cumin seed/ Jeera	59 Turmeric/ Haldi	60 Zedoary/ Gandhmul
XI	*Antilithics*				green						
XII	*Antiparasitic*							red			
XIII	*Antiphlogistics*	orange								orange	
XIV	*Antiscorbutics*										
XV	*Antiseptic*	blue									
XVI	*Antispasmodic*			yellow			yellow				
XVII	*Aphrodisiac*										
XVIII	*Aromatic*			green	green					green	green
XIX	*Astringent*	tan	tan	tan					tan		
XX	*Bitters and Bitter tonics*	purple									

Plants/Name/ Uses	61 Lemon grass/ Bhustrina	62 Dhub grass/ Dhub	63 Nut grass/ Nagar motha	64 Indian Rhubarb/ Nasatara	65 Indian thornapple/ Dhatura	66 Horse Gram/ Kulthi	67 Trailing Eclipta/ Bhringaraja	68 Lesser cardamom/ Choti elachi	69 Embelia	70 Indian gooseberry/ Amla
XI Antilithics		■ (green)				■ (green)				■ (green)
XII Antiparasitic										
XIII Antiphlogistics					■ (peach)		■ (peach)			
XIV Antiscorbutics										
XV Antiseptic										
XVI Antispasmodic	■ (yellow)				■ (yellow)		■ (yellow)			
XVII Aphrodisiac					■ (dark blue)					
XVIII Aromatic					■ (green)			■ (green)		
XIX Astringent		■ (tan)	■ (tan)		■ (tan)	■ (tan)				■ (tan)
XX Bitters and Bitter tonics										■ (purple)

Plants/Name/ Uses		71 Wood apple/ Kaith	72 Asafoetida/ Hing	73 Indian Banyan/ Bargat	74 Indian fig tree/ Gular	75 Peepal Tree/ Pipal	76 Indian Coffee Plum/ Talispatri	77 Fennel/ Barisauf	78 Chirata, Indian Gentian/ Bhuchiretta	79 Liquorice/ Mulahatti	80 Kashmir Tree/ Gumbhar
XI	*Antilithics*				■ (green)						
XII	*Antiparasitic*										
XIII	*Antiphlogistics*				■ (light peach)						
XIV	*Antiscorbutics*	■ (orange)			■ (orange)						
XV	*Antiseptic*										
XVI	*Antispasmodic*		■ (yellow)								
XVII	*Aphrodisiac*							■ (navy)			
XVIII	*Aromatic*	■ (green)									
XIX	*Astringent*	■ (grey)		■ (grey)	■ (grey)	■ (grey)	■ (grey)				
XX	*Bitters and Bitter tonics*								■ (purple)		

Plants/Name/ Uses		81 Levant cotton/ Kapaas	82 Chaulmugra seeds	83 East Indian screw Tree/ Mamorphali	84 Indian Sarsaparilla/ Anatmul	85 Lady Finger/ Bhindi	86 Kurchi	87 Chaul moogra	88 Hyssop	89 Indian indigo/ Nil	90 Common Jasmine/ Chameli
XI	Antilithics	■ (green)			■ (green)		■ (green)			■ (green)	
XII	Antiparasitic		■ (red)								
XIII	Antiphlogistics		■ (orange)		■ (orange)						
XIV	Antiscorbutics										
XV	Antiseptic										
XVI	Antispasmodic						■ (yellow)				
XVII	Aphrodisiac	■ (dark)									
XVIII	Aromatic										
XIX	Astringent			■ (tan)			■ (tan)				■ (tan)
XX	Bitters and Bitter tonics						■ (purple)				

Plants/Name/ Uses		91 Arabian Jasmine/ Moghra	92 Purging nut/ Jamalgota	93 Walnut/ Akhrot	94 Henna plant/ Mahendi	95 Garden Cress/ Chandrasur	96 Flax/ Linseed Alsi	97 Mohwa Tree/ Mahua	98 Indian Kamila/ Kamila	99 Mango Tree/ Aam	100 Wild Chamomile/ Babunah
Antilithics	XI										
Antiparasitic	XII										
Antiphlogistics	XIII										
Antiscorbutics	XIV										
Antiseptic	XV										
Antispasmodic	XVI										
Aphrodisiac	XVII										
Aromatic	XVIII										
Astringent	XIX										
Bitters and Bitter tonics	XX										

Plants/Name/Uses		101 Indian lilac	102 Yellow champa/ Champake	103 Sensitive plant/ Laifavanti	104 Bitter Gourd/ Karela	105 Indian Mulberry/ Shatoot	106 Horse raddish Tree/ Sajana	107 Cowhage/ Kavatch	108 Curry-leaf Tree/ Katnimb	109 Banana/ Kela	110 Sacred Lotus/ Kamal
XI	Antilithics	green		green			green				
XII	Antiparasitic	brown									
XIII	Antiphlogistics		peach	peach			peach				
XIV	Antiscorbutics				orange		orange			orange	
XV	Antiseptic										
XVI	Antispasmodic	yellow					yellow				
XVII	Aphrodisiac						dark blue	dark blue		dark blue	
XVIII	Aromatic					olive					
XIX	Astringent		gray					gray		gray	gray
XX	Bitters and Bitter tonics	purple	purple								

Plants/Name / Uses	111 Oleander/ Kanaer	112 Black Cumin/ Kalajiri	113 Coral Jasmine/ Harsinghar	114 Holy Basil/ Tulsi	115 Prickly pear/ Chappal	116 Indian Trumpet flower/Jagdala	117 Yellow oxalis/ Chukatripatti	118 Bada gokhru	119 Date Palm/ Khajur	120 Longleaf Indian pine/ Chir
XI Antilithics								■ (green)		
XII Antiparasitic										
XIII Antiphlogistics	■ (peach)									
XIV Antiscorbutics							■ (orange)			
XV Antiseptic				■ (cyan)						■ (cyan)
XVI Antispasmodic								■ (yellow)		
XVII Aphrodisiac								■ (navy)	■ (navy)	
XVIII Aromatic				■ (green)						■ (green)
XIX Astringent			■ (tan)			■ (tan)				
XX Bitters and Bitter tonics			■ (purple)							

Plants/Name/Uses	121 Betel Leaf vine/Pan	122 Long pepper Pippli	123 Black pepper/Kali mirch	124 Ispaghula/Isabgul	125 Pagoda Tree/Temple Tree/Champa	126 Indian beech/Karanj	127 Purslane/Kurfa	128 Babchi seeds/Bakuchi	129 Red Sandal wood/Rakta-chandan	130 Pomegranate/Anar
XI Antilithics										■ (green)
XII Antiparasitic						■ (brown)				
XIII Antiphlogistics	■ (orange)			■ (orange)		■ (orange)			■ (orange)	
XIV Antiscorbutics							■ (orange)			
XV Antiseptic	■ (blue)	■ (blue)				■ (blue)				
XVI Antispasmodic										
XVII Aphrodisiac	■ (navy)							■ (navy)		
XVIII Aromatic	■ (green)		■ (green)							
XIX Astringent	■ (grey)			■ (grey)	■ (grey)	■ (grey)	■ (grey)		■ (grey)	■ (grey)
XX Bitters and Bitter tonics		■ (purple)								

	Plants/Name/Uses	131 Emetic nut/ Mainphala	132 Radish/ Muli	133 Indian Snakeroot/ Chota chand	134 Castor oil plant/ Erand	135 Indian madder/ Manjith	136 Silk Cotton Tree/ Shemul	137 Tooth brush tree/ Pilu	138 Sandalwood Tree/Safed Chandan	139 Three-leaf soapberry/ Ritha	140 Ashoka tree/ Ashoka
XI	Antilithics		green				green				
XII	Antiparasitic									brown	
XIII	Antiphlogistics				peach			peach			
XIV	Antiscorbutics	orange	red					red			
XV	Antiseptic								cyan		
XVI	Antispasmodic	yellow				yellow					
XVII	Aphrodisiac	navy					navy				
XVIII	Aromatic							green			
XIX	Astringent	grey				grey			grey	grey	grey
XX	Bitters and Bitter tonics							purple	purple		

Plants/Name/ Uses		141 Costus Root/ Kuth	142 Marking nut tree	143 Sesame/ Til	144 Shala tree/ Sala	145 Country Mallow Bala	146 Black nightshade/ Makoy	147 Indian Red Wood Tree/ Rohini	148 Poison nut tree/ Kuchala	149 Clearingnut tree/ Nirmali	150 Lodh Tree/ Lodh
XI	Antilithics										
XII	Antiparasitic	■ (brown)									
XIII	Antiphlogistics				■ (light orange)			■ (orange)			
XIV	Antiscorbutics					■ (orange-red)					
XV	Antiseptic	■ (blue)	■ (blue)								
XVI	Antispasmodic	■ (yellow)	■ (yellow)								
XVII	Aphrodisiac	■ (dark blue)			■ (dark blue)	■ (dark blue)			■ (dark blue)		
XVIII	Aromatic	■ (green)									
XIX	Astringent	■ (grey)			■ (grey)	■ (grey)		■ (grey)	■ (grey)		■ (grey)
XX	Bitters and Bitter tonics								■ (purple)		

	Plants/Name/Uses	151 Jambolan/ Jamun	152 Tamarind/ Imli	153 Teak tree/ Sagvan	154 Arjun Tree/ Arjuna	155 Beleric myrobalan/ Bahera	156 Chebulic myrobalan/ Hirda	157 Tulip tree/ Paras	158 Heart-leaved Moonseed/ Amruta	159 Bishop's weed/ Ajowan	160 Small caltrops/ Chota gokhru
XI	*Antilithics*				green						green
XII	*Antiparasitic*										
XIII	*Antiphlogistics*	orange									
XIV	*Antiscorbutics*										
XV	*Antiseptic*		blue								
XVI	*Antispasmodic*						yellow			yellow	
XVII	*Aphrodisiac*								purple		purple
XVIII	*Aromatic*										
XIX	*Astringent*	gray	gray	gray	gray	gray	gray	gray			gray
XX	*Bitters and Bitter tonics*										

Plants/Name/Uses	161 Fenugreek/Methi	162 Wheat/Gehun	163 Indian squill/Jangle piyaz	164 Stinging nettle/Bichu	165 Indian copal tree/Dhup	166 Indian Valerian/Jata-manasi	167 Wild cumin/Kali Zeera	168 Common violet/Banaf shah	169 Winter cherry/Asvagandha	170 Ginger/Sonth	171 Indian Date/Ber
XI Antilithics			■ (green)	■ (green)							
XII Antiparasitic					■ (brown)		■ (brown)		■ (brown)		
XIII Antiphlogistics	■ (peach)	■ (peach)							■ (peach)		
XIV Antiscorbutics	■ (orange)										
XV Antiseptic											
XVI Antispasmodic				■ (yellow)		■ (yellow)					
XVII Aphrodisiac	■ (navy)								■ (navy)		
XVIII Aromatic	■ (green)					■ (green)				■ (green)	
XIX Astringent	■ (tan)			■ (tan)			■ (tan)	■ (tan)			■ (tan)
XX Bitters and Bitter tonics											■ (purple)

		1	2	3	4	5	6	7	8	9	10
	Plants/Name/ Uses	Catechu Tree	Rough-chaff/ Latjira	Sweet flag/ Vacha	Malabar nut/Vasaka	Bael	Garlic/ Lahsun	Aloe/Ghee Kunvar	Siamese Ginger/ Kulanjan	Devil's Tree/ Chatian	Cholai-bhajee
XXI	*Carminative*			green			green		green		
XXII	*Demulcent*					orange					orange
XXIII	*Dentifrices*										
XXIV	*Deobstruents*										
XXV	*Depuratives*		yellow								
XXVI	*Diaphoretic*						dark blue				
XXVII	*Discutient*						olive green				
XXVIII	*Diuretic*		tan	tan			tan				tan
XXIX	*Emetic*		purple	purple							

Plants/ Name/Uses		11 Cashew nut/ Kaju	12 Fish berry/ Kakamari	13 The creat/ Kirata	14 Custard apple/ Sharifa	15 Celery/ Ajmoda	16 Areca nut/ Sufari	17 Prickly Poppy/ Bramha dana	18 Elephant Creeper/ Samudra-palaka	19 Indian Birthwort/ Isharmul	20 Wormwood/ Afsanthin
XXI	*Carminative*			green		green	green			green	
XXII	*Demulcent*							cream			
XXIII	*Dentifrices*						orange			blue	
XXIV	*Deobstruents*										
XXV	*Depuratives*										
XXVI	*Diaphoretic*										
XXVII	*Discutient*										olive
XXVIII	*Diuretic*	grey				grey		grey	grey		
XXIX	*Emetic*							purple			

Plants/Name/Uses	21 Santonica, Wormseed/ Kirmala	22 Indian Wormwood/ Nagadamani	23 Shatavari	24 Indiana Belladonna/ Agnurshera	25 Nim Tree/ Nim	26 Thyme-leaved Gratiola/ Bamb	27 Indian bamboo/ Bans	28 Indian Oak/ Sumudraphal	29 Variegated Bauhinia/ Kachanar	30 Barberry/ Rasaut
XXI Carminative					■ (green)					
XXII Demulcent			■ (orange)				■ (orange)			
XXIII Dentifrices										
XXIV Deobstruents	■ (blue)	■ (blue)								
XXV Depuratives									■ (yellow)	■ (yellow)
XXVI Diaphoretic										■ (navy)
XXVII Discutient					■ (green)					
XXVIII Diuretic			■ (khaki)		■ (khaki)	■ (khaki)				
XXIX Emetic					■ (purple)			■ (purple)		

Plants/Name/Uses	31 Pigweed/Punarnava	32 Toddy Palm/Tad	33 Life plant (Zakhme-hayata)	34 Flame of the forest/Palas	35 Fever nut/Katkaranja	36 Calendula/Zergut	37 Dillo oil Tree/Champa	38 Swallow wart/Madar	39 Indian Hemp/Bhang, Charas	40 Papaya/Papita
XXI *Carminative*										green
XXII *Demulcent*										
XXIII *Dentifrices*		orange								
XXIV *Deobstruents*										
XXV *Depuratives*				yellow			blue			
XXVI *Diaphoretic*	dark					dark		dark		
XXVII *Discutient*					green				green	
XXVIII *Diuretic*				tan				tan	tan	
XXIX *Emetic*							purple	purple		

Plants/Name/Uses		41 East Indian senna/Senna	42 Purging cassia/Amaltas	43 Senna Sophera/Kasunda	44 White silk cotton Tree/Sveta Salmali	45 Indian Pennywort/Brahmi	46 Cinnamon/Dalchini	47 Pareira brava/Path	48 Butterfly pea/Aparajita	49 Coconut Palm/Nariyal	50 Guggul
XXI	*Carminative*						■ (green)				■ (green)
XXII	*Demulcent*				■ (orange)				■ (orange)		■ (orange)
XXIII	*Dentifrices*									■ (orange-red)	
XXIV	*Deobstruents*										
XXV	*Depuratives*				■ (yellow)	■ (yellow)		■ (yellow)			
XXVI	*Diaphoretic*										
XXVII	*Discutient*										■ (navy)
XXVIII	*Diuretic*				■ (olive)	■ (olive)	■ (olive)	■ (olive)			
XXIX	*Emetic*					■ (purple)			■ (purple)		

Plants/Name/Uses		51 Jute Plant/ Kost	52 Indian Cherry/ Lasora	53 Coriander/ Dhania	54 Garlic pear/ Varuna	55 Poison bulb/ Kanwal	56 Saffron crocus/ Kesar	57 Purging croton/ Jamalgota	58 Cumin seed/Jeera	59 Turmeric/ Haldi	60 Zedoary/ Gandhmul
XXI	Carminative	■		■			■		■	■	■
XXII	Demulcent	■	■		■						
XXIII	Dentifrices										
XXIV	Deobstruents										
XXV	Depuratives										■
XXVI	Diaphoretic					■		■			
XXVII	Discutient							■	■		
XXVIII	Diuretic			■				■	■		■
XXIX	Emetic					■		■			

	Plants/Name/Uses	61 Lemon grass/ Bhustrina	62 Dhub grass/ Dhub	63 Nut grass/ Nagar motha	64 Indian Rhubarb/ Nasatara	65 Indian thornapple/ Dhatura	66 Horse Gram/ Kulthi	67 Bhringaraja	68 Lesser cardamom/ Choti elachi	69 Embelia	70 Indian gooseberry/ Amla
XXI	*Carminative*	■ (green)				■ (green)			■ (green)		
XXII	*Demulcent*										
XXIII	*Dentifrices*										
XXIV	*Deobstruents*							■ (cyan)			
XXV	*Depuratives*										■ (yellow)
XXVI	*Diaphoretic*			■ (navy)							
XXVII	*Discutient*					■ (olive)		■ (olive)			
XXVIII	*Diuretic*	■ (tan)		■ (tan)			■ (tan)		■ (tan)		■ (tan)
XXIX	*Emetic*							■ (purple)			

Plants/Name/Uses		71 Wood apple/ Kaith	72 Asafoetida Hing	73 Indian Banyon/ Bargat	74 Indian fig tree/ Gular	75 Peepal tree/ Pipal	76 Indian Coffee Plum/ Talispatri	77 Fennel/ Barisauf	78 Chirata/ Indian Gentian Bhuchiretta	79 Liquorice/ Mulahatti	80 Kashmir Tree/ Gumbhar
XXI	Carminative	■	■		■		■				
XXII	Demulcent	■								■	■
XXIII	Dentifrices										
XXIV	Deobstruents										
XXV	Depuratives										
XXVI	Diaphoretic		■	■			■				
XXVII	Discutient				■						
XXVIII	Diuretic		■					■			
XXIX	Emetic										

Plants/Name/Uses	81 Levant Cotton/ Kapaas	82 Chaulmugra seeds	83 East Indian Screw tree/ Mamorphali	84 Indian Sarsaparilla/ Anatmul	85 Lady Finger/ Bhindi	86 Kurchi	87 Chaulmoogra	88 Hyssop	89 Indian indigo/Nil	90 common Jasmine/ Chameli
XXI *Carminative*						██		██		
XXII *Demulcent*	██		██	██	██					
XXIII *Dentifrices*										
XXIV *Deobstruents*								██	██	
XXV *Depuratives*								██		
XXVI *Diaphoretic*				██				██		
XXVII *Discutient*										
XXVIII *Diuretic*				██	██			██		
XXIX *Emetic*										

	Plants/Name/Uses	91 Arabian Jasmine/moghra	92 Purging nut/Jamalgota	93 Walnut/akhrot	94 Henna plant/Mahendi	95 Garden Cress/Chandrasur	96 Flax/Linseed/Alsi	97 Mohwa Tree/Mahua	98 Indian Kamila/Kamila	99 Mango Tree/Aam	100 Wild Chamomile/Babunah
XXI	Carminative										■ (green)
XXII	Demulcent						■ (orange)	■ (orange)			
XXIII	Dentifrices										
XXIV	Deobstruents	■ (blue)									
XXV	Depuratives					■ (yellow)					
XXVI	Diaphoretic										
XXVII	Discutient		■ (green)								■ (olive)
XXVIII	Diuretic	■ (tan)				■ (tan)	■ (tan)			■ (tan)	
XXIX	Emetic							■ (purple)			■ (purple)

	Plants/Name/ Uses	101 Indian lilac	102 Yellow champa/ Champake	103 Sensitive plant/ Laifavanti	104 Bitter Gourd/ Karela	105 Indian Mulberry/ Shatoot	106 Horse raddish Tree/Sajana	107 Cowhage/ Kavatch	108 Curry-leaf Tree/ Katnimb	109 Banana/ Kela	110 Sacred Lotus/ Kamal
XXI	Carminative		■ (green)								
XXII	Demulcent		■ (orange)							■ (orange)	■ (orange)
XXIII	Dentifrices										
XXIV	Deobstruents	■ (blue)				■ (blue)					
XXVI	Depuratives										
XXVII	Diaphoretic										
XXVII	Discutient	■ (green)					■ (green)				
XXVIII	Diuretic	■ (grey)	■ (grey)				■ (grey)	■ (grey)		■ (grey)	■ (grey)
XXIX	Emetic			■ (purple)	■ (purple)	■ (purple)	■ (purple)				

Plants/Name/Uses		111 Oleander/ Kanaer	112 Black Cumin/ Kalajira	113 Coral Jasmine/ Harsinghar	114 Holy Basil/ Tulsi	115 Prickly pear/ Chappal	116 Indian Trumpet flower/ Jagdala	117 Yellow oxalis/ Chukatripatti	118 Bada gokhru	119 Date Palm/ Khajur	120 Longleaf Indian pine/Chir
XXI	Carminative		■		■		■				
XXII	Demulcent				■	■	■		■	■	
XXIII	Dentifrices										
XXIV	Deobstruents										
XXV	Depuratives										
XXVI	Diaphoretic			■	■		■				■
XXVII	Discutient	■					■				
XXVIII	Diuretic		■	■	■				■	■	
XXIX	Emetic										

		121	122	123	124	125	126	127	128	129	130
	Plants /Name/ Uses	Betel Leaf vine/Pan	Long pepper/ Pippli	Black pepper/ Kali mirch	Ispaghula/ Isabgul	Pagoda Tree/ Temple Tree/ Champa	Indian beech/ Karanj	Purslane/ Kurfa	Babchi seeds/ Bakuchi	Red Sandal wood/ Rakta-chandan	Pomegranate/ Anar
XXI	*Carminative*	■	■	■			■				
XXII	*Demulcent*				■						
XXIII	*Dentifrices*										
XXIV	*Deobstruents*			■					■		
XXV	*Depuratives*										
XXVI	*Diaphoretic*								■	■	
XXVII	*Discutient*	■		■							
XXVIII	*Diuretic*		■	■	■			■			
XXIX	*Emetic*										

Plants/Name/ Uses		131 Emetic nut/ Mainphala	132 Radish/ Muli	133 Indian snakeroot/ Chota chand	134 Castor oil plant/ Erand	135 Indian madder/ Manjith	136 Silk Cotton Tree/Shemul	137 Tooth brush tree Pilu	138 Sandal Wood Tree/ Safed Chandan	139 Three-leaf soapberry/ Ritha	140 Ashoka tree/ Ashoka
XXI	*Carminative*	green	green			green		green			
XXII	*Demulcent*						orange				
XXIII	*Dentifrices*										
XXIV	*Deobstruents*					blue		blue	blue		
XXV	*Depuratives*										
XXVI	*Diaphoretic*	dark purple				dark purple					
XXVII	*Discutient*					green			green	green	
XXVIII	*Diuretic*	grey	grey						grey		
XXIX	*Emetic*	purple			purple		purple			purple	

Plants/Name/ Uses	141 Costus Root/Kuth	142 Marking nut tree	143 Sesame/ Til	144 Shala tree/Sala	145 Country Mallow/Bala	146 Black nightshade/ Makoy	147 Indian Red Wood Tree/ Rohini	148 Poison nut tree/ Kuchala	149 Clearingnut tree Nirmali	150 Lodh Tree/ Lodh
XXI Carminative										
XXII Demulcent			orange		orange				orange	
XXIII Dentifrices										
XXIV Deobstruents										
XXV Depuratives	yellow									
XXVI Diaphoretic			navy	navy		navy				
XXVII Discutient										
XXVIII Diuretic	grey		grey			grey	grey			
XXIX Emetic										

Plants/Name/Uses		151 Jambolan/ Jamun	152 Tamarind/ Imli	153 Teak tree/ Sagvan	154 Arjun Tree/ Arjuna	155 Beleric myrobalan/ Bahera	156 Chebulic myrobalan/ Hirda	157 Tulip tree/ Paras	158 Heart-leaved Moonseed/ Amruta	159 Bishop's weed/ Ajowan	160 Small caltrops/ Chota gokhru
XXI	*Carminative*	■	■								
XXII	*Demulcent*										■
XXIII	*Dentifrices*						■				
XXIV	*Deobstruents*										
XXV	*Depuratives*										
XXVI	*Diaphoretic*		■	■							
XXVII	*Discutient*	■									
XXVIII	*Diuretic*										
XXIX	*Emetic*										

Plants/Name/Uses	161 Fenugreek/Methi	162 Wheat/Gehun	163 Indian squill/Jangle piyaz	164 Nettle Stinging nettle/Bichu	165 Indian copal tree/Dhup	166 Indian Valerian/Jata-manasi	167 Wild cumin/Kali Zeera	168 Violet/Banaf shah	169 Winter cherry/Asvagandha	170 Ginger/Sonth	171 Indian Date/Ber
XXI Carminative	green					green	green			green	
XXII Demulcent	orange							orange			
XXIII Dentifrices											
XXIV Deobstruents			blue			blue			blue		
XXV Depuratives											yellow
XXVI Diaphoretic											
XXVII Discutient	green						green	navy			
XXVIII Diuretic	tan		tan	tan		tan	tan	tan	tan	tan	
XXIX Emetic			purple					purple			

Plants/Name/Uses	1 Catechu Tree	2 Rough-chaff/Latjira	3 Sweet flag/Vacha	4 Malabar nut/Vasaka	5 Bael	6 Garlic/Lahson	7 Aloe/Ghee Kunvar	8 Siamese Ginger/Kulanjan	9 Devil's Tree/Chatian	10 Cholai-bhajee
XXX Emmenagogues			■ (green)			■ (green)	■ (green)		■ (green)	
XXXI Emollients		■ (brown)								
XXXII Expectorants		■ (peach)	■ (peach)	■ (peach)		■ (peach)				
XXXIII Eyedrops and lotions		■ (orange)			■ (orange)					
XXXIV Febrifuge				■ (blue)	■ (blue)	■ (blue)				
XXXV Galactagogue									■ (yellow)	■ (yellow)
XXXVI Gargles										
XXXVII Hair tonic										
XXXVIII Liniments			■ (olive)			■ (olive)				
XXXIX Narcotic										

Plants/Name/Uses		11 Cashew nut/ Kaju	12 Fish berry/ Kakamari	13 The creat/ Kirata	14 Custard apple/ Sharifa	15 Celery/ Ajmoda	16 Areca nut/ Sufari	17 Prickly Poppy/ Bramha danai	18 Elephant Creeper/ Samudra-palaka	19 Indian Birthwort/ Isharmul	20 Wormwood/ Afsanthin
Emmenagogues	XXX									■ (green)	■ (green)
Emollients	XXXI	■ (brown)									
Expectorants	XXXII							■ (peach)			
Eyedrops and lotions	XXXIII						■ (orange)	■ (orange)			
Febrifuge	XXXIV			■ (cyan)						■ (cyan)	
Galactagogue	XXXV										
Gargles	XXXVI										
Hair tonic	XXXVII				■ (green)						
Liniments	XXXVIII						■ (tan)				
Narcotic	XXXIX		■ (purple)					■ (purple)			

	21	22	23	24	25	26	27	28	29	30
Plants/Name/Uses	Santonica, Wormseed/ Kirmala	Indian Wormwood/ Nagadamani	Shatavari	Indiana Belladona/ Agnurshera	Nim Tree/ Nim	Thyme-leaved Gratiola/ Bamb	Indian bamboo/ Bans	Indian Oak/ Sumudraphal	Variegated Bauhinia/ Kachanara	Barberry/ Rasaut
XXX *Emmenagogues*		■ (green)			■ (green)		■ (green)			
XXXI *Emollients*					■ (red)					
XXXII *Expectorants*		■ (orange)	■ (orange)		■ (orange)		■ (orange)	■ (orange)		
XXXIII *Eyedrops and lotions*										
XXXIV *Febrifuge*					■ (blue)	■ (blue)	■ (blue)	■ (blue)		
XXXV *Galactagogue*			■ (yellow)							
XXXVI *Gargles*										
XXXVII *Hair tonic*					■ (green)					
XXXVIII *Liniments*					■ (gray)	■ (gray)				
XXXIX *Narcotic*										

Plants/Name/Uses		31 Pigweed/ Punarnava	32 Toddy Palm/ Tad	33 Life plant/ Zakhme-hayat	34 Flame of the forest/ Palas	35 Fever nut Katkararnj	36 Calendula/ Zergut	37 Dilo oil Tree/ Champa	38 Swallow wart/ Madar	39 Indian Hemp/ Bhang, Charas	40 Papaya/ Papita
Emmenagogues	XXX	●				●					●
Emollients	XXXI					●					
Expectarants	XXXII								●		
Eyedrops and lotions	XXXIII	●					●	●			
Febrifuge	XXXIV					●			●		●
Galactagogue	XXXV						●				
Gargles	XXXVI		●								
Hair tonic	XXXVII									●	
Liniments	XXXVIII					●		●			
Narcotic	XXXIX									●	

Plants/Name/Uses		41 East Indian senna/ Senna	42 Purging cassia/ Amaltas	43 Senna Sophera/ Kasunda	44 White silk cotton Tree/ Sveta Salmali	45 Indian Pennywort/ Brahmi	46 Cinnamon/ Dalchini	47 Pareira brava/ Path	48 Butterfly pea/ Aparajita	49 Coconut Palm/ Nariyal	50 Guggul
XXX	Emmenagogues										■
XXXI	Emollients		■		■						
XXXII	Expectorants			■							■
XXXIII	Eyedrops and lotions								■		
XXXIV	Febrifuge		■		■	■					
XXXV	Galactagogue										
XXXVI	Gargles									■	■
XXXVII	Hair tonic									■	
XXXVIII	Liniments						■				
XXXIX	Narcotic										

	51 Jute Plant/Kost	52 Indian Cherry/Lasora	53 Coriander/Dhania	54 Garlic pear/Varuna	55 Poison bulb/Kanwal	56 Saffron crocus/Kesar	57 Purging croton/Jamalgota	58 Cumin seed/Jeera	59 Turmeric/Haldi	60 Zedoary/Gandhmul
XXX *Emmenagogues*								●		
XXXI *Emollients*	●		●							
XXXII *Expectorants*										●
XXXIII *Eyedrops and lotions*			●						●	
XXXIV *Febrifuge*				●	●					●
XXXV *Galactagogue*								●		
XXXVI *Gargles*		●	●							
XXXVII *Hair tonic*							●			
XXXVIII *Liniments*				●			●			
XXXIX *Narcotic*						●				

Plants/Name/Uses	61 Lemon grass/ Bhustrina	62 Dhub grass/ Dhub	63 Nut grass/ Nagar motha	64 Indian Rhubarb/ Nasatara	65 Indian thornapple/ Dhatura	66 Horse Gram/ Kulthi	67 Trailing Eclipta/ Bhringaraja	68 Lesser cardamom/ Choti elachi	69 Embelia	70 Indian gooseberry/ Amla
XXX Emmenagogues			■ (green)			■ (green)		■ (green)		
XXXI Emollients										■ (brown)
XXXII Expectorants										
XXXIII Eyedrops and lotions										■ (orange)
XXXIV Febrifuge						■ (blue)				■ (blue)
XXXV Galactagogue			■ (yellow)							
XXXVI Gargles									■ (navy)	■ (navy)
XXXVII Hair tonic					■ (olive green)		■ (olive green)			
XXXVIII Liniments					■ (tan)					
XXXIX Narcotic					■ (purple)					

Plants/Name/Uses		71 Wood apple/ Kaith	72 Asafoetida/ Hing	73 Indian Banyan/ Bargat	74 Indian fig tree/ Gular	75 Peepal tree/ Pipal	76 Indian Coffee Plum/ Talispatri	77 Fennel/ Barisauf	78 Chirata, Indian Gentian/ Bhuchiretta	79 Liquorice/ Mulahatti	80 Kashmir Tree/ Gumbhar
XXX	*Emmenagogues*		green								green
XXXI	*Emollients*										
XXXII	*Expectorants*		orange				orange			orange	
XXXIII	*Eyedrops and lotions*										
XXXIV	*Febrifuge*								blue		
XXXV	*Galactagogue*							yellow			yellow
XXXVI	*Gargles*				dark	dark	dark				
XXXVII	*Hair tonic*										
XXXVIII	*Liniments*									olive	
XXXIX	*Narcotic*										

Plants/Name/Uses		81 Levant/ Kapaas	82 Chaulmugra seeds	83 East Indian screw Tree/ Mamorphali	84 Indian Sarsaparilla/ Anatmul	85 Lady Finger/ Bhindi	86 Kurchi/ Kurchi	87 Chaulmoogra	88 Hyssop	89 Indian indigo/ Nil	90 Common Jasmine/ Chambeli
Emmenagogues	XXX	green							green		
Emollients	XXXI					brown					
Expectorants	XXXII	peach					peach		peach		
Eyedrops and lotions	XXXIII				orange				orange		orange
Febrifuge	XXXIV	blue				blue	blue				
Galactagogue	XXXV	yellow									
Gargles	XXXVI						navy		navy		
Hair tonic	XXXVII										
Liniments	XXXVIII	olive	olive						olive	olive	
Narcotic	XXXIX										

Plants/Name/Uses		91 Arabian Jasmine/ Moghra	92 Purging nut/ Jamalgota	93 Walnut/ Akhrot	94 Henna plant/ Mahendi	95 Garden Cress/ Chandrasur	96 Flax/ Linseed/ Alsi	97 Mohwa Tree/ Mahua	98 Indian Kamila/ Kamila	99 Mango Tree/ Aam	100 Wild Chamomile/ Babunah
XXX	*Emmenagogues*	green			green						green
XXXI	*Emollients*										
XXXII	*Expectorants*							peach			
XXXIII	*Eyedrops and lotions*										
XXXIV	*Febrifuge*		cyan								cyan
XXXV	*Galactagogue*		yellow			yellow					
XXXVI	*Gargles*		dark navy	dark navy	dark navy			dark navy		dark navy	
XXXVII	*Hair tonic*				olive green						
XXXVIII	*Liniments*		tan				tan				
XXXIX	*Narcotic*										

Plants/Name/Uses		101 Indian lilac	102 Yellow champa/Champake	103 Sensitive plant/Laifavanti	104 Bitter Gourd/Karela	105 Indian Mulberry/Shatoot	106 Horse raddish Tree/Sajana	107 Cowhage/Kavatch	108 Curry-leaf Tree/Katnimb	109 Banana/Kela	110 Sacred Lotus/Kamal
XXX	Emmenagogues	green	green			green		green			
XXXI	Emollients								brown		
XXXII	Expectorants		light orange								light orange
XXXIII	Eyedrops and lotions		orange-red								
XXXIV	Febrifuge		blue			blue	blue		blue		
XXXV	Galactagogue				yellow						
XXXVI	Gargles						navy			navy	
XXXVII	Hair tonic										
XXXVIII	Liniments		khaki		khaki		khaki				
XXXIX	Narcotic										

Plants/Name/Uses	111 Oleander/ Kanaer	112 Black Cumin/ Kalajira	113 Coral Jasmine/ Harsinghar	114 Holy Basil/ Tulsi	115 Prickly pear/ Chappal	116 Indian Trumpet flower/Jagdala	117 Yellow oxalis/ Chukatripatti	118 Bada gokhru	119 Date Palm/ Khajur	120 Longleaf Indian pine/Chir
XXX *Emmenagogues*		■ green						■ green		
XXXI *Emollients*				■ brown						
XXXII *Expectorants*			■ orange	■ orange						
XXXIII *Eyedrops and lotions*	■ orange								■ orange	
XXXIV *Febrifuge*			■ blue	■ blue					■ blue	
XXXV *Galactagogue*		■ yellow								
XXXVI *Gargles*								■ navy		
XXXVII *Hair tonic*										
XXXVIII *Liniments*										
XXXIX *Narcotic*							■ purple			

Plants/Name/Uses		121 Betel Leaf vine/Pan	122 Long pepper/Pippli	123 Black pepper/Kali mirch	124 Ispaghula/Isabgul	125 Pagoda Tree/Temple Tree/Champa	126 Indian beech/Karanj	127 Purslane/Kurfa	128 Babchi seeds/Bakuchi	129 Red Sandal wood/Rakta-chandan	130 Pomegranate/Anar
XXX	Emmenagogues		green	green							green
XXXI	Emollients		brown		brown			brown		brown	
XXXII	Expectorants										
XXXIII	Eyedrops and lotions	orange								orange	orange
XXXIV	Febrifuge					blue	blue				blue
XXXV	Galactagogue										
XXXVI	Gargles	navy		navy							navy
XXXVII	Hair tonic			olive							olive
XXXVIII	Liniments						khaki				
XXXIX	Narcotic										

Plants/Name/Uses		131 Emetic nut/ Mainphala	132 Radish/ Muli	133 Indian snakeroot/ Chota chand	134 Castor oil plant/ Erand	135 Indian madder/ Manjith	136 Silk Cotton Tree/ Shemul	137 Tooth brush tree/Pilu	138 Sandal Wood Tree/Safed Chandan	139 Three-leaf soapberry/ Ritha	140 Ashoka tree/ Ashok
XXX	Emmenagogues	green	green	green	green			green			
XXXI	Emollients				brown				brown		
XXXII	Expectorants	tan				tan				tan	
XXXIII	Eyedrops and lotions			orange	orange		orange				
XXXIV	Febrifuge		blue	blue	blue						
XXXV	Galactagogue				yellow						
XXXVI	Gargles										
XXXVII	Hair tonic										
XXXVIII	Liniments										
XXXIX	Narcotic										

Plants/Name/Uses		141 Costus Root/Kuth	142 Marking nut tree	143 Sesame/Til	144 Shala tree/Sala	145 Country Mallow/Bala	146 Black nightshade/Makoy	147 Indian Red Wood Tree/Rohini	148 Poison nut tree/Kuchala	149 Clearingnut tree/Nirmali	150 Lodh Tree/Lodh
XXX	Emmenagogues			■ (green)							
XXXI	Emollients			■ (brown)							
XXXII	Expectorants	■ (peach)					■ (peach)				
XXXIII	Eyedrops and lotions					■ (orange)				■ (orange)	■ (orange)
XXXIV	Febrifuge					■ (blue)		■ (blue)	■ (blue)		
XXXV	Galactagogue			■ (yellow)							
XXXVI	Gargles						■ (purple)	■ (purple)			■ (purple)
XXXVII	Hair tonic	■ (olive)		■ (olive)							
XXXVIII	Liniments										
XXXIX	Narcotic										

	Plants/Name/Uses	151 Jambolan/ Jamun	152 Tamarind/ Imli	153 Teak tree/ Sagvan	154 Arjun Tree/ Arjuna	155 Beleric myrobalan/ Bahera	156 Chebulic myrobalan/ Hirda	157 Tulip tree/ Paras	158 Heart-leaved Moonseed/ Amruta	159 Bishop's weed/ Ajowan	160 Small caltrops/ Chota gokhru
XXX	*Emmenagogues*										
XXXI	*Emollients*			■							
XXXII	*Expectorants*										
XXXIII	*Eyedrops and lotions*		■	■		■	■				
XXXIV	*Febrifuge*		■		■						
XXXV	*Galactagogue*										
XXXVI	*Gargles*	■					■				■
XXXVII	*Hair tonic*			■		■					
XXXVIII	*Liniments*		■			■					
XXXIX	*Narcotic*										

Plants/Name/Uses		161 Fenugreek/ Methi	162 Wheat/ Gehun	163 Indian squill/ Jangle piyaz	164 Nettle Stinging nettle/ Bichu	165 Indian copal tree/ Dhup	166 Indian Valerian/ Jata-manasi	167 Wild cumin/ Kali Zeera	168 Violet/ Banaf shah	169 Winter cherry/ Asvagandha	170 Ginger/ Sonth	171 Indian Date/Ber
Emmenagogues	XXX	green	green	green	green		green				green	
Emollients	XXXI	brown	brown						brown			
Expectorants	XXXII			light orange								
Eyedrops and lotions	XXXIII										orange	orange
Febrifuge	XXXIV								blue	blue	blue	
Galactagogue	XXXV	yellow										yellow
Gargles	XXXVI											
Hair tonic	XXXVII	olive green					olive green					
Liniments	XXXVIII											
Narcotic	XXXIX										purple	

Plants/Name/Uses		1 Catechu Tree	2 Rough-chaff/ Latjira	3 Sweet flag/Vacha	4 Malabar nut/Vasaka	5 Bael	6 Garlic/ Lahson	7 Aloe/Ghee Kunvar	8 Siamese Ginger/Kulanjan	9 Devil's Tree/ Chatian	10 Cholai-bhajee
Pectoral	XL	green	green	green	green		green		green		
Purgative	XLI		brown		brown			brown		brown	
Refrigerant	XLII										
Rheumatism	XLIII		orange	orange	orange		orange	orange	orange		
Rubefacient	XLIV						blue				
Sedative	XLV		yellow								
(Remedies) Skin diseases	XLVI	navy	navy		navy		navy		navy		navy
Stomachics	XLVII		olive			olive	olive		olive		
Styptics	XLVIII	olive-gray									
Suppurative	XLIX										
Tonic	L	blue		blue			blue	blue	blue		

Plants/Name/Uses	11 Cashew nut/ Kaju	12 Fish berry/ Kakamari	13 The creat/ Kirata	14 Custard apple/ Sharifa	15 Celery/ Ajmoda	16 Areca nut/ Sufari	17 Prickly Poppy/ Bramha danai	18 Elephant Creeper/ Samudra-palaka	19 Indian Birthwort/ Isharmul	20 Wormwood/ Afsanthin
XL *Pectoral*							green			
XLI *Purgative*				brown	brown	brown	brown			
XLII *Refrigerant*										
XLIII *Rheumatism*	orange					orange				orange
XLIV *Rubefacient*	blue									
XLV *Sedative*										
XLVI *(Remedies) Skin diseases*	dark	dark					dark		dark	dark
XLVII *Stomachics*			green				green			green
XLVIII *Styptics*										
XLIX *Suppurative*				purple			purple	purple		
L *Tonic*					light blue				light blue	light blue

	Plants/Name/Uses	21 Santonica, Wormseed/ Kirmala	22 Indian Wormwood/ Nagadamani	23 Shatavari	24 Indiana Belladonna/ Agnurshera	25 Neem Tree	26 Thyme-leaved Gratiola	27 Indian bamboo/ Bans	28 Indian Oak/ Sumudra-phal	29 Variegated Bauhinia/ Kachanara	30 Barberry/ Rasaut
XL	*Pectoral*		green				green	green	green		
XLI	*Purgative*					brown	brown				
XLII	*Refrigerant*			light orange		light orange					
XLIII	*Rheumatism*			orange	orange		orange				
XLIV	*Rubefacient*										
XLV	*Sedative*				yellow						
XLVI	*(Remedies) Skin diseases*					navy		navy		navy	navy
XLVII	*Stomachics*	green	green			green		green			green
XLVIII	*Styptics*		olive								
XLIX	*Suppurative*										
L	*Tonic*	light blue		light blue		light blue					light blue

Plants/Name/Uses		31 Pigweed/ Punarnava	32 Toddy Palm/ Tad	33 Life plant/ ZakhmeHayat	34 Flame of the forest/ Palas	35 Fever nut/ Katkaranja	36 Calendula/ Zergut	37 Dilo oil Tree/ Champa	38 Swallow wart/ Madar	39 Indian Hemp/ Bhang, Charas	40 Papaya/ Papita
Pectoral	XL										
Purgative	XLI										
Refrigerant	XLII										
Rheumatism	XLIII										
Rubefacient	XLIV										
Sedative	XLV										
(Remedies) Skin diseases	XLVI										
Stomachics	XLVII										
Styptics	XLVIII										
Suppurative	XLIX										
Tonic	L										

Plants/Name/Uses		41 East Indian senna/ Senna	42 Purging cassia/ Amaltas	43 Senna Sophera/ Kasunda	44 White silk cotton Tree/ Sveta Salmali	45 Indian Pennywort/ Brahmi	46 Cinnamon/ Dalchini	47 Pareira brava/ Path	48 Butterfly pea/ Aparajita	49 Coconut Palm/ Nariyal	50 Guggul
XL	*Pectoral*										
XLI	*Purgative*	brown	brown	brown	brown	brown			brown	brown	brown
XLII	*Refrigerant*									cream	
XLIII	*Rheumatism*	orange				orange	orange				orange
XLIV	*Rubefacient*									cyan	
XLV	*Sedative*						yellow	yellow			
XLVI	*(Remedies) Skin diseases*	dark blue		dark blue		dark blue		dark blue		dark blue	dark blue
XLVII	*Stomachics*							green			green
XLVIII	*Styptics*				grey					grey	green
XLIX	*Suppurcative*										
L	*Tonic*	light blue				light blue	light blue	light blue			light blue

Plants/Name/Uses		51 Jute Plant/ Kost	52 Indian Cherry/ Lasora	53 Coriander/ Dhania	54 Garlic pear/ Varuna	55 Poison bulb/ Kanwal	56 Saffron crocus/ Kesar	57 Purging croton/ Jamalgota	58 Cumin seed/ Jeera	59 Turmeric/ Haldi	60 Zedoary/ Gandhmul
Pectoral	XL		green				green				green
Purgative	XLI	brown	brown			brown		brown			
Refrigerant	XLII			orange							
Rheumatism	XLIII			orange	orange		orange				
Rubefacient	XLIV				blue	blue		blue			
Sedative	XLV				yellow		yellow				
(Remedies) Skin diseases	XLVI	dark blue								dark blue	
Stomachics	XLVII	green			green		green		green		green
Styptics	XLVIII									khaki	
Suppurative	XLIX										
Tonic	L	light blue	light blue	light blue	light blue	light blue	light blue	light blue	light blue	light blue	light blue

Plants/Name/Uses		61 Lemon grass/ Bhustrina	62 Dhub grass/ Dhub	63 Nut grass/ Nagar motha	64 Indian Rhubarb/ Nasatara	65 Indian thornapple Dhatura	66 Horse Gram/ Kulthi	67 Trailing Eclipta/ Bhringaraja	68 Lesser cardamom/ Choti elachi	69 Embelia	70 Indian gooseberry/ Amla
XL	*Pectoral*							green		green	
XLI	*Purgative*				brown			brown			
XLII	*Refrigerant*	orange							orange		orange
XLIII	*Rheumatism*	orange			orange	orange		orange			
XLIV	*Rubefacient*	blue									
XLV	*Sedative*					yellow					
XLVI	*(Remedies) Skin diseases*				navy	navy		navy		navy	navy
XLVII	*Stomachics*	green		green		green				green	green
XLVIII	*Styptics*		khaki					khaki			
XLIX	*Suppurative*										
L	*Tonic*	light blue					light blue			light blue	light blue

Plants/Name/Uses		71 Wood apple/Kaith	72 Asafoetida/Hing	73 Indian Banyan/Bargat	74 Indian fig tree/Gular	75 Peepal tree/Pipal	76 Indian Coffee Plum/Talispatri	77 Fennel/Barisauf	78 Chirata, Indian Gentian/Bhuchiretta	79 Liquorice/Mulahatti	80 Kashmir Tree/Gumbhar
Pectoral	XL						green			green	
Purgative	XLI		brown			brown			brown	brown	brown
Refrigerant	XLII										
Rheumatism	XLIII			orange	orange						
Rubefacient	XLIV										
Sedative	XLV		yellow								
(Remedies) Skin diseases	XLVI		navy			navy			navy		
Stomachics	XLVII	green			green	green	green		green		green
Styptics	XLVIII										
Suppurative	XLIX					purple					
Tonic	L	light blue	light blue	light blue	light blue		light blue		light blue		light blue

		81	82	83	84	85	86	87	88	89	90
	Plants/Name/Uses	Levant cotton/ Kapaas	Chaulmugra seeds	East Indian screw Tree/ Mamorphali	Indian Sarsaparilla/ Anatmul	Lady Finger/ Bhindi	Kurchi/ Kurchi	Chaulmoogra	Hyssop	Indian indigo/Nil	Common Jasmine/ Chambeli
XL	*Pectoral*	green								green	green
XLI	*Purgative*	brown							brown		
XLII	*Refrigerant*										
XLIII	*Rheumatism*		orange		orange						
XLIV	*Rubefacient*										
XLV	*Sedative*	yellow									
XLVI	*(Remedies) Skin diseases*	dark	dark	dark	dark	dark		dark		dark	dark
XLVII	*Stomachics*						green		green		
XLVIII	*Styptics*										
XLIX	*Suppurative*	purple									
L	*Tonic*	light blue	light blue		light blue		light blue		light blue		

Plants/Name/Uses	91 Arabian Jasmine/ Moghra	92 Purging nut/ Jamalgota	93 Walnut/ Akhrot	94 Henna plant/ Mahendi	95 Garden Cress/ Chandrasur	96 Flax/ Linseed/ Alsi	97 Mohwa Tree/ Mahua	98 Indian Kamila/ Kamila	99 Mango Tree/ Aam	100 Wild Chamomile/ Babunah
XL Pectoral					green	green			green	
XLI Purgative		brown	brown		brown	brown	brown	brown	brown	
XLII Refrigerant										
XLIII Rheumatism		orange		orange	orange		orange			orange
XLIV Rubefacient		blue			blue					
XLV Sedative										yellow
XLVI (Remedies) Skin diseases		dark	dark	dark	dark	dark	dark	dark	dark	
XLVII Stomachics										
XLVIII Styptics		tan				tan			tan	
XLIX Suppurative	purple					purple				
L Tonic		light blue				light blue	light blue		light blue	light blue

Plants/Name/Uses		101 Indian lilac	102 Yellow champa/Champake	103 Sensitive plant/Laifavanti	104 Bitter Gourd/Karela	105 Indian Mulberry/Shatoot	106 Horse raddish Tree/Sajana	107 Cowhage/Kavatch	108 Curry-leaf Tree/Katnimb	109 Banana/Kela	110 Sacred Lotus/Kamal
XL	*Pectoral*						●				●
XLI	*Purgative*	●	●		●	●	●		●	●	
XLII	*Refrigerant*										●
XLIII	*Rheumatism*		●		●	●	●	●			
XLIV	*Rubefacient*						●				
XLV	*Sedative*										
XLVI	*(Remedies) Skin diseases*	●			●				●		●
XLVII	*Stomachics*		●		●						
XLVIII	*Styptics*									●	●
XLIX	*Suppurative*						●				
L	*Tonic*		●	●	●	●	●	●		●	●

Plants/Name/Uses	111 Oleander/ Kanaer	112 Black Cumin/ Kalajiri	113 Coral Jasmine/ Harsinghar	114 Holy Basil/ Tulsi	115 Prickly pear/ Chappal	116 Indian Trumpet flower/Jagdala	117 Yellow oxalis/ Chukatripatti	118 Bada gokhru	119 Date Palm/ Khajur	120 Longleaf Indian pine/Chir
XL *Pectoral*				green	green				green	
XLI *Purgative*			brown		brown				brown	
XLII *Refrigerant*					tan		tan		tan	
XLIII *Rheumatism*			orange			orange				
XLIV *Rubefacient*										
XLV *Sedative*										
XLVI *(Remedies) Skin diseases*	purple		purple	purple	purple					purple
XLVII *Stomachics*				green		green	green			green
XLVIII *Styptics*										
XLIX *Suppurative*					purple		purple			
L *Tonic*		blue	blue	blue		blue			blue	

		121	122	123	124	125	126	127	128	129	130
Plants/Name/Uses		Betel Leaf vine/Pan	Long pepper/Pippli	Black pepper/Kali mirch	Ispaghula/Isabgul	Pagoda Tree/Temple Tree/Champa	Indian beech/Karanj	Purslane/Kurfa	Babchi seeds/Bakuchi	Red Sandal wood/Rakta-chandan	Pomegranate/Anar
XL	*Pectoral*	●	●	●	●	●	●				
XLI	*Purgative*		●	●	●	●			●		
XLII	*Refrigerant*										●
XLIII	*Rheumatism*		●		●	●	●				
XLIV	*Rubefacient*	●	●	●		●					
XLV	*Sedative*										
XLVI	*(Remedies) Skin diseases*			●		●	●	●	●	●	
XLVII	*Stomachics*		●	●			●		●	●	●
XLVIII	*Styptics*							●			●
XLIX	*Suppurative*										
L	*Tonic*	●	●	●	●	●	●		●	●	●

Plants/Name/Uses		131 Emetic nut/ Mainphala	132 Radish/ Muli	133 Indian snakeroot/ Chota chand	134 Castor oil plant/ Erand	135 Indian madder/ Manjith	136 Silk Cotton Tree/ Shemul	137 Tooth brush tree/Pilu	138 Sandal Wood Tree/ Safed Chandan	139 Three-leaf soapberry/ Ritha	140 Ashoka tree/ Ashoka
XL	*Pectoral*	green			green						
XLI	*Purgative*	brown	brown		brown					brown	
XLII	*Refrigerant*										
XLIII	*Rheumatism*	orange			orange			orange			
XLIV	*Rubefacient*										
XLV	*Sedative*	yellow	yellow			yellow			yellow		yellow
XLVI	*(Remedies) Skin diseases*			navy	navy	navy	navy		navy		
XLVII	*Stomachics*				olive						
XLVIII	*Styptics*						khaki				
XLIX	*Suppurative*			purple	purple						
L	*Tonic*	light blue	light blue			light blue			light blue		light blue

Plants/Name/Uses		141 Costus Root/Kuth	142 Marking nut tree	143 Sesame/Til	144 Shala tree/Sala	145 Country Mallow/Bala	146 Black nightshade/Makoy	147 Indian Red Wood Tree/Rohini	148 Poison nut tree/Kuchala	149 Clearingnut tree/Nirmali	150 Lodh Tree/Lodh
XL	*Pectoral*		green	green							
XLI	*Purgative*				brown		brown	brown				brown
XLII	*Refrigerant*											
XLIII	*Rheumatism*			orange		orange	orange		orange	orange		
XLIV	*Rubefacient*			blue								
XLV	*Sedative*		yellow	yellow			yellow	yellow				
XLVI	*(Remedies) Skin diseases*		navy	navy	navy			navy		navy		
XLVII	*Stomachics*		olive				olive			olive		olive
XLVIII	*Styptics*											
XLIX	*Suppurative*						purple				purple	
L	*Tonic*		light blue				light blue	light blue	light blue	light blue		

Plants/Name/Uses	151 Jambolan/ Jamun	152 Tamarind/ Imli	153 Teak tree/ Sagvan	154 Arjun Tree/ Arjuna	155 Beleric myrobalan/ Bahera	156 Chebulic myrobalan/ Hirda	157 Tulip tree/ Paras	158 Heart-leaved Moonseed/ Amruta	159 Bishop's weed/ Ajowan	160 Small caltrops/ Chota gokhru
XL *Pectoral*				green	green					green
XLI *Purgative*		brown			brown	brown				brown
XLII *Refrigerant*		peach	peach							peach
XLIII *Rheumatism*		orange						orange		
XLIV *Rubefacient*										
XLV *Sedative*			yellow							
XLVI *(Remedies) Skin diseases*		navy	navy				navy	navy		
XLVII *Stomachics*	olive					olive		olive		
XLVIII *Styptics*										
XLIX *Suppurative*										
L *Tonic*	light blue	light blue		light blue	light blue	light blue	light blue	light blue	light blue	light blue

Plants/Name/Uses		161 Fenugreek/ Methi	162 Wheat/ Gehun	163 Indian squill/ Jangle piyaz	164 Nettle Stinging nettle/ Bichu	165 Indian copal tree/ Dhup	166 Indian Valerian/ Jata-manasi	167 Wild cumin/ Kali Zeera	168 Violet/ Banaf shah	169 Winter cherry/ Asvagandha	170 Ginger/ Sonth	171 Indian date/ Ber
XL	*Pectoral*	green		green	green		green			green	green	green
XLI	*Purgative*			brown	brown		brown		brown			brown
XLII	*Refrigerant*		peach									
XLIII	*Rheumatism*	orange		orange	orange						orange	orange
XLIV	*Rubefacient*				cyan						cyan	
XLV	*Sedative*						yellow					
XLVI	*(Remedies) Skin diseases*		navy	navy	navy	navy		navy		navy		
XLVII	*Stomachics*						olive	olive			olive	
XLVIII	*Styptics*				gray							gray
XLIX	*Suppurative*											
L	*Tonic*	lt.blue		lt.blue						lt.blue	lt.blue	lt.blue

Plants/Name/Uses	1 Catechu Tree	2 Rough-chaff/Latjira	3 sweet flag/Vacha	4 Malabarnut/Vasaka	5 Bael	6 Garlic/Lason	7 Aloe/Ghee Kunvar	8 Siamese Ginger/Kulanjan	9 Devil's Tree/Chatian	10 Cholai bhajee
LI Urinary system diseases								■		
LII Uterine diseases	■	■	■							■
LIII Venereal diseases	■	■		■						■
LIV Vesicant						■				
LV Vulneraries										

Plants/Name/Uses		11 Cashew nut/Kaju	12 Fish berry/Kakamari	13 The creat/Kirata	14 Custard apple/Sharifa	15 Celery/Ajmoda	16 Areca nut/Sufari	17 Prickly Poppy/Bramha danai	18 Elephant Creeper/Samudra-palaka	19 Indian Birthwort/Isharmul	20 Wormwood/Afsanthin
LI	Urinary system diseases										
LII	Uterine diseases	■ (brown)					■ (green)				■ (brown)
LIII	Venereal diseases							■ (tan)			
LIV	Vesicant	■ (orange)									
LV	Vulneraries										

Plants/Name/Uses		21 Santoruca Wormseed/ Kirmala	22 Indian Wormwood/ Nagadamani	23 Shatavari	24 Indiana Belladona/ Agnurshera	25 Nim Tree/ Nim	26 Thyme-leaved/ Gratiola/ Bamb.	27 Indian bamboo/ Bans	28 Indian Oak/ Sumudraphal	29 Variegated Bauhinia/ Kachanara	30 Barberry/ Rasaut
LI	Urinary system diseases		(green)								
LII	Uterine diseases		(brown)								
LIII	Venereal diseases								(brown)	(tan)	
LIV	Vesicant										
LV	Vulneraries										

Plants/Name/Uses		31 Pigweed/ Punarnava	32 Toddy Palm/ Tad	33 Life plant Zakhme- hayata	34 Flame of the forest/ Palas	35 Fever nut/ Katkararnja	36 Calendula/ Zergut	37 Dilo oil Tree/ Champa	38 Swallow wart/ Madar	39 Indian Hemp/ Bhang, Charas	40 Papaya/ Papita
LI	*Urinary system diseases*		■ (green)					■ (green)			
LII	*Uterine diseases*					■ (brown)					
LIII	*Venereal diseases*	■ (peach)	■ (peach)					■ (peach)	■ (peach)		
LIV	*Vesicant*		■ (orange)							■ (blue)	■ (orange)
LV	*Vulneraries*					■ (blue)	■ (blue)	■ (blue)	■ (blue)	■ (blue)	

Plants/Name/Uses		41 Indian senna/ Senna	42 Purgining cassia/ Amaltas	43 Senna Sophera/ Kasunda	44 White silk cotton Tree/ Sveta Salmali	45 Indian Pennywort/ Brahmi	46 Cinnamon/ Dalchini	47 Pareira brava/ path	48 Butterfly pea/ Aparajita	49 Coconut Palm/ Nariyal	50 Guggul
LI	*Urinary system diseases*										
LII	*Uterine diseases*						▨				▨
LIII	*Venereal diseases*			▨	▨	▨			▨	▨	
LIV	*Vesicant*										
LV	*Vulneraries*										

Plants/Name/Uses		51 Jute Plant/ Kost	52 Indian Cherry/ Lasora	53 Coriander/ Dhania	54 Garlic pear/ Varuna	55 Poison bulb/ Kanwal	56 Saffron crocus/ Kesar	57 Purging croton/ Jamalgota	58 Cumin seed/ Jeera	59 Turmeric/ Haldi	60 Zedoary/ Gandhmul
LI	*Urinary system diseases*	🟩			🟩					🟩	
LII	*Uterine diseases*		🟫								
LIII	*Venereal diseases*								🟧	🟧	🟧
LIV	*Vesicant*				🟧						
LV	*Vulneraries*										

Plants/Name/Uses	61 Lemon grass/ Bhustrina	62 Dhub grass/ Dhub	63 Nut grass/ Nagar motha	64 Indian Rhubarb/ Nasatara	65 Indian thornapple/ Dhatura	66 Horse Gram/ Kulthi	67 Trailing Eclipta Bhringaraja	68 Lesser cardamom/ Choti elachi	69 Embelia	70 Indian gooseberry/ Amla
LI *Urinary system diseases*							(green)			
LII *Uterine diseases*										
LIII *Venereal diseases*		(orange)					(orange)			(orange)
LIV *Vesicant*			(blue)							
LV *Vulneraries*										

Plants/Name/Uses		71 Wood apple/ Kaith	72 Asafoetida/ Hing	73 Indian Banyan/ Bargat	74 Indian fig tree/ Gular	75 Peepal tree/ Pipal	76 Indian Coffee Plum/ Talispatri	77 Fennel/ Barisauf	78 Chirata, Indian Gentian/ Bhuchiretta	79 Liquorice/ Mulahatti	80 Kashmir Tree/ Gumbhar
LI	*Urinary system diseases*									(green)	(green)
LII	*Uterine diseases*										
LIII	*Venereal diseases*				(orange)	(orange)				(orange)	(orange)
LIV	*Vesicant*										
LV	*Vulneraries*										

Plants/Name/Uses		81 Levant cotton/ Kapaas	82 Chaulmugra seeds	83 East Indian screw Tree/ Mamorphali	84 Indian Sarsaparillo/ Anatmul	85 Lady Finger Bhindi	86 Kurchi/ Kurchi	87 Chaulmoogra	88 Hyssop	89 Indian indigo/ Nil	90 Common Jasmine/ Chambeli
LI	*Urinary system diseases*						green			green	green
LII	*Uterine diseases*								brown		
LIII	*Venereal diseases*		orange		orange						
LIV	*Vesicant*	blue									
LV	*Vulneraries*								blue	blue	

		91	92	93	94	95	96	97	98	99	100
	Plants/Name/Uses	Arabian Jasmine/ Moghra	Purging nut/ Jamalgota	Walnut Akhrot	Henna plant/ Mahendi	Garden Cress/ Chandrasur	Flax/ Linseed/ Alsi	Mohwa Tree/ Mahua	Indian Kamila/ Kamila	Mango Tree/ Aam	Wild Chamomile/ Babunah
LI	Urinary system diseases				(green)						
LII	Uterine diseases				(brown)						
LIII	Venereal diseases			(orange)		(orange)	(orange)			(orange)	
LIV	Vesicant										
LV	Vulneraries				(blue)		(blue)				

Plants/Name/Uses	101 Indian lilac	102 Yellow champa/Champake	103 Sensitive plant/Laifavanti	104 Bitter Gourd/Karela	105 Indian Mulberry/Shatoot	106 Horse raddish Tree/Sajana	107 Cowhage/Kavatch	108 Curry-leaf Tree/Katnimb	109 Banana/Kela	110 Sacred Lotus/Kamal
LI *Urinary system diseases*										
LII *Uterine diseases*									▓	
LIII *Venereal diseases*		▓				▓			▓	
LIV *Vesicant*							▓			
LV *Vulneraries*					▓	▓				

Plants/Name/Uses		111 Oleander/Kanaer	112 Black Cumin/Kalajiri	113 Coral Jasmine/Harsinghar	114 Holy Basil/Tulsi	115 Prickly pear/Chappal	116 Indian Trumpet flower/Jagdala	117 Yellow oxalis/Chukatripatti	118 Bada gokhru	119 Date Palm/Khajur	120 Long leaf Indian pine/Chir
LI	*Urinary system diseases*										
LII	*Uterine diseases*								▨		
LIII	*Venereal diseases*					▨			▨	▨	▨
LIV	*Vesicant*										
LV	*Vulneraries*										

Plants/Name/Uses	121 Betal Leaf vine/ Pan	122 Long pepper/ Pippli	123 Black pepper/ Kali mirch	124 Ispaghula/ Isabgul	125 Pagoda Tree/Temple Tree/ Champa	126 Indian beech/ Karanj	127 Purslane/ Kurfa	128 Babchi seeds/ Bakuchi	129 Red Sandal wood/ Raktachandan	130 Pomegranate/ Anar
LI *Urinary system diseases*										(green)
LII *Uterine diseases*										(brown)
LIII *Venereal diseases*		(orange)	(orange)	(orange)	(orange)	(orange)				
LIV *Vesicant*										
LV *Vulneraries*										

Plants/Name/Uses		131 Emetic nut/ Mainphala	132 Radish/ Muli	133 Indian snakeroot/ Chota chand	134 Castor oil plant/ Erand	135 Indian madder/ Manjith	136 Silk Cotton Tree/ Shemul	137 Tooth brush tree/Pilu	138 Sandal Wood Tree/ Safed Chandan	139 Three-leaf soapberry/ Ritha	140 Ashoka tree/ Ashoka
LI	*Urinary system diseases*					▨	▨				
LII	*Uterine diseases*					▨					▨
LIII	*Venereal diseases*	▨	▨				▨		▨		
LIV	*Vesicant*							▨			
LV	*Vulneraries*			▨	▨						

Plants/Name/Uses		141 Costus Root/ Kuth	142 Marking nut tree	143 Sesame/ Til	144 Shala tree/ Sala	145 Country Mallow/Bala	146 Indian nightshade/ Makoy	147 Indian Red Wood Tree/ Rohini	148 Poison nut tree/ Kuchala	149 Clearingnut tree/Nirmali	150 Lodh Tree/ Lodh
LI	Urinary system diseases										
LII	Uterine diseases										▓
LIII	Venereal diseases		▨		▨	▨	▨			▨	
LIV	Vesicant		▧								
LV	Vulneraries					▨					

Plants/Name/Uses		151 Jambolan/ Jamun	152 Tamarind/ Imli	153 Teak tree/ Sagvan	154 Arjun Tree/ Arjuna	155 Beleric myrobalan/ Bahera	156 Chebulic myrobalan/ Hirda	157 Tulip tree/ Paras	158 Heart-leaved Moonseed/ Amruta	159 Bishop's weed/ Ajowan	160 Small calthrops/ Chota gokhru
LI	*Urinary system diseases*		green	green			green		green		
LII	*Uterine diseases*										
LIII	*Venereal diseases*							orange	orange		orange
LIV	*Vesicant*				blue						
LV	*Vulneraries*										

Plants/Name/Uses		161 Fenugreek/ Methi	162 Wheat/ Gehun	163 Indian squill/ Jangle piyaz	164 Nettle Stinging nettle/ Bichu	165 Indian copal tree/ Dhup	166 Indian Valerian/ Jata-manasi	167 Wild cumin/ Kali Zeera	168 Violet Banaf shah	169 Winter cherry/ Asvagandha	170 Ginger/ Sonth	171 Indian Date/Ber
LI	Urinary system diseases											
LII	Uterine diseases		▉							▉		
LIII	Venereal diseases					▉						▉
LIV	Vesicant											
LV	Vulneraries					▉						

Appendix-1

Plant	Botanical name of the plants (family)	Common name
1.	Acacia catechu (Leguminosae)	Catechu tree
2.	Achyranthes aspera Linn. (Amaranthaceae)	Rough-chaff
3.	Acorus calamus Linn. (Araceae)	Sweet flag
4.	Adhatoda vasica Nees (Acanthaceae)	Malabar nut
5.	Aegle marmelos Corr. (Rutaceae)	Bael
6.	Allium sativum Linn. (Liliaceae)	Garlic
7.	Aloe barbadensis Mill (Liliaceae)	Aloe
8.	Alpinia galangal Willd. (Scitamineae)	Siamese ginger
9.	Alstonia scholaris R.Br. (Apocynaceae)	Devil tree
10.	Amaranthus polygamous Linn. (Amaranthaceae)	Cholaibhajee
11.	Anacardium occidentale Linn. (Anacardiaceae)	Cashew nut
12.	Anamirta cocculus (Menispermaceae)	Fish berry
13.	Andrographis paniculata Nees. (Acanthaccae)	Creat
14.	Annona squamosa Linn. (Annonaceae)	Custard apple

(*Continued*)

225

(Continued)

Plant	Botanical name of the plants (family)	Common name
15.	Apium graveolens Linn. (Umbelliferae)	Celery
16.	Areca catechu Linn. (Palmacea)	Areca nut
17.	Argemone mexicana Linn (Papaveraceae)	Prickly poppy
18.	Argyreia speciosa sweet. (Convolvulaceae)	Elephant creeper
19.	Aristolochia indica Linn.(Aristolochiaceae)	Indian birthwort
20.	Artemisia absinthum Linn. (Asteraceae)	Wormwood
21.	Artemisia maritima Linn. (Compositae)	Santonica, Wormseed
22.	Artemisia nilagirica (Asteraceae)	Indian wormwood
23.	Asparagus racemosus willd. (Liliaceae)	Shatavari
24.	Atropa acuminata Royle ex Lindley (Solanaceae)	Indian belladonna
25.	Azadirachta indica Adr. Juss. (Meliaceae)	Nim tree
26.	Bacopa monnieri Linn. Pennell (Scrophulariaceae)	Thyme-leaved gratiola
27.	Bambusa bambos Voss. (Gramineae)	Indian bamboo
28.	Barringtonia acutangula Gaertn. (Lecythidaceae)	Indian oak
29.	bauhinia variegata Linn. (Leguminosae)	Variegated bauhinia
30.	Berberis aristata DC. (Berberidaceae)	Barberry
31.	Boerhavia diffusa Linn. (Nyctaginaceae)	Pigweed
32.	Borassus flabellifer Linn. (Palmae)	Toddy palm

(Continued)

(Continued)

Plant	Botanical name of the plants (family)	Common name
33.	Bryophyllum pinnatum Kurz. (Crassulaceae)	Life plant
34.	Butea monosperma Kuntz.(Leguminosae)	Flame of the forest
35.	Caesalpinia crista Lam. (Leguminosae)	Fever nut
36.	Calendula officinalis Linn. (Asteraceae)	Calendula
37.	Calophyllum inophyllum Linn. (Guttiferae)	Dilo oil tree
38.	Calotropis procera R.Br. (Apocynaceae)	Swallow wort
39.	Cannabis sativa Linn. (Cannabaceae)	Indian hemp
40.	Carica papaya. Linn. (Caricaceae)	Papaya
41.	Cassia angustifolia Vahl. (Leguminosae)	East Indian senna
42.	Cassia fistula Linn. (Leguminosae)	Purging cassia
43.	Cassia sophera Linn. (Leguminosae)	Senna sophera
44.	Ceiba pentandra Gaertn. (Malvaceae)	White silk cotton tree
45.	Centella asiatica Urban. (Umbelliferae)	Indian Pennywort
46.	Cinnamomum zeylanicum Breyn. (Lauraceae)	Cinnamon
47.	Cissampelos pareira Linn. (Mcnispermaceae)	Pareira brava
48.	Clitoria ternatea Linn. (Leguminosae)	Butterfly pea
49.	Cocos nucifera Linn. (Arecaceae)	Coconut palm

(Continued)

(*Continued*)

Plant	Botanical name of the plants (family)	Common name
50.	Commiphora mukul Engl. (Burseraceae)	Guggul
51.	Corchorus olitorius Linn. (Tiliaceae)	Jute plant
52.	Cordia dichotoma Forst f. (Boraginaceae)	Indian cherry
53.	Coriandrum sativum Linn. (Umbelliferae)	Coriander
54.	Crataeva nurvala Buch-Ham. (Capparaceae)	Garlic pear
55.	Crinum asiaticum Linn. (Amaryllidaceae)	Poison bulb
56.	Crocus sativus Linn. (Iridaceae)	Saffron crocus
57.	Croton tinglium Linn. (Euphorbiaceae)	Purging croton
58.	Cuminum cyminum Linn. (Umbelliferae)	Cumin seed
59.	Curcuma longa Linn. (Zingiberaceae)	Turmeric
60.	Curcuma zedoaria Rosc. (Zingiberaceae)	Zedoary
61.	Cymbopogon citratus Stapf. (Gramineae)	Lemon grass
62.	Cynodon dactylon Pers. (Gramineae)	Dhub grass
63.	Cyperus rotundus Linn. (Cyperaceae)	Nut grass
64.	Damera turpethum (Convolvulaceae)	Indian rhubarb
65.	Datura metel Linn. (Solanaceae)	Indian thornapple
66.	Dolichos biflorus Linn. (Papileonaceae)	Horse Gram
67.	Eclipta alba Hassk (Asteraceae)	Trailing eclipta (bhringraj)
68.	Elettaria cardamomum Maton. (Zingiberaceae)	Lesser cardamom
69.	Embelia ribes Burm. (Primulaceae)	Embelia
70.	Emblica officinalis Gaertn. (Euphorbiaceae)	Indian gooseberry

(*Continued*)

(*Continued*)

Plant	Botanical name of the plants (family)	Common name
71.	Feronia limonia Swingle. (Rutaceae)	Wood apple
72.	Ferula narthex Boiss (Umbellifarae)	Asafoetida
73.	Ficus benghalensis Linn. (Moraceae)	Indian banyan
74.	Ficus glomerata Roxb. (Moraceae)	Indian fig tree
75.	Ficus religiosa Linn. (Moraceae)	Peepal tree
76.	Flacourtia jangomas Raesch. (Salicaceae)	Indian coffee Plum
77.	Foeniculum vulgare Mill. (Umbelliferae)	Fennel
78.	Gentian kurroo Royle (Gentianaceae)	Chirata, Indian Gentian
79.	Glycyrrhiza glabra Linn. (Leguminosae)	Liquorice
80.	Gmelina arborea Roxb. (Lamiaceae)	Kashmir tree
81.	Gossypium herbaceum Linn. (Malvaceae)	Levant cotton
82.	Gynocardia odorata R.Br. (Achariaceae)	Chaulmugra seeds
83.	Helicteres isora Linn. (Malvaceae)	East Indian screw tree
84.	Hemidesmus indicus R.Br. (Apocynaceae)	Indian sarsaparilla
85.	Hibiscus esculentus Linn. (Malvaceae)	Lady finger
86.	Holarrhena antidysenterica Wall. (Apocynaceae)	Kurchi
87.	Hydnocarpus kurzii Warb (Flacourtiaceae)	Chaulmoogra
88.	Hyssopus officinalis Linn. (Labiatae)	Hyssop
89.	Indigofera tinctoria Linn. (Leguminosae)	Indian indigo
90.	Jasminum officinale forma grandiflorum Kobusk (Oleaceae)	Common jasmine

(*Continued*)

(Continued)

Plant	Botanical name of the plants (family)	Common name
91.	jasminum sambac. Aiton. (Oleaceae)	Arabian Jasmine
92.	Jatropha curcas Linn. (Euphorbiaceae)	Purging nut
93.	Juglans regia Linn. (Juglandaceae)	Walnut
94.	Lawsonia inermis Linn. (Lythraceae)	Henna plant
95.	Lepidium sativum Linn. (Cruciferae)	Garden cress
96.	Linum usitatissimum Linn. (Linaceae)	Flax/Linseed
97.	Madhuca indica Gmel. (Sapotaceae)	Mohwa tree
98.	Mallotus oppositifolius. (Euphorbiaceae)	Indian kamila
99.	Mangifera indica (Anacardiaceae)	Mango tree
100.	Matricaria chamomilla Linn. (Compositae)	Wild chamomile
101.	Melia azedarach Linn. (Meliaceae)	Indian lilac
102.	Magnolia champaca Linn. (Magnoliaceae)	Yellow champa
103.	Mimosa pudica Linn. (Leguminosae)	Sensitive plant
104.	Momordica charantia Linn. (Cucurbitaceae)	Bitter gourd
105.	Morinda tinctoria Roxb. (Rubiaceae)	Indian mulberry
106.	Moringa oleifera Lam. (Moringaceae)	Horseraddish Tree
107.	Mucuna pruriens (Leguminosae)	Cowhage
108.	Murraya koenigii spreng. (Rutaceae)	Curry-leaf Tree
109.	Musa paradisiaca Linn. (Musaceae)	Banana
110.	Nelumbium speciosum Willd. (Nelumbonaceae)	Sacred lotus
111.	Nerium indicum Mill. (Apocynaceae)	Oleander
112.	Nigella sativa Linn. (Ranunculaceae)	Black cumin
113.	Nyctanthes arbor-tristis Linn. (Oleaceae)	Coral jasmine
114.	Ocimum sanctum Linn. (Lamiaceae)	Holy basil

(Continued)

(*Continued*)

Plant	Botanical name of the plants (family)	Common name
115.	Opuntia dillenii Haw. (Cactaceae)	Prickly pear
116.	Oroxylum indicum Vent. (Bignoniaceae)	Indian trumpet flower
117.	Oxalis stricta Linn. (Oxalidaceae)	Yellow wood sorrel
118.	Pedalium murex Linn. (Pedaliaceae)	Bada gokhru
119.	Phoenix dactylifera Linn. (Arecaceae)	Date palm
120.	Pinus roxburghii Sargent. (Coniferae)	Longleaf Indian pine
121.	Piper betel Linn. (piperaceae)	Betel vine
122.	Piper longum Linn. (Piperaceae)	Long pepper
123.	Piper nigrum Linn. (Piperaceae)	Black pepper
124.	Plantago ovata Forsk. (Plantaginaceae)	Ispaghula
125.	Plumeria rubra Linn. Var Acutifolia bailey (Apocynaceae)	Pagoda tree/ Temple tree
126.	Pongamia pinnata (L.) Pierre (Leguminosae)	Indian beech tree
127.	Portulaca Oleracea Linn. (Portulacaceae)	Purslane
128.	Psoralea corylifolia Linn. (Leguminosae)	Babchi seeds
129.	Pterocarpus santalinus Linn. (Leguminosae)	Red Sandalwood
130.	Punica granatum Linn. (Punicaceae)	Pomegranate
131.	Randia dumetorum Lamk. (Rubiaceae)	Emetic nut tree
132.	Raphanus sativus Linn. (Brassicaceae)	Radish
133.	Rauvolfia serpentina Berth. (Apocynaceae)	Indian snakeroot
134.	Ricinus communis Linn. (Euphorbiaceae)	Castor oil plant
135.	Rubia cordifolia Linn. (Rubiaceae)	Indian madder
136.	Salmalia malabarica Schott and Endl. (Malvaceae)	Silk Cotton Tree

(*Continued*)

(Continued)

Plant	Botanical name of the plants (family)	Common name
137.	Salvadora persica Linn. (Salvadoraceae)	Toothbrush tree
138.	Santalum album Linn. (Santalaceae)	Sandalwood tree
139.	Sapindus trifoliatus Linn. (Sapindaceae)	Three-leaf soapberry
140.	Saraca asoca Linn. (Leguminosae)	Ashoka tree
141.	Saussurea lappa Clarke. (Compositae)	Costus root
142.	Semecarpus anacardium Linn. f. (Anacardiaceae)	Marking nut tree
143.	Sesamum indicum Linn. (Pedaliaceae)	Sesame
144.	Shorea robusta Gaertn. F. (Dipterocarpaceae)	Shala tree
145.	Sida cordifolia Linn. (Malvaceae)	Country Mallow
146.	Solanum nigrum Linn. (Solanaceae)	Black nightshade
147.	Soymida febrifuga Adr. Juss. (Meliaceae)	Indian red wood tree
148.	Strychnos nux-vomica Linn. (Loganiaceae)	Poison nut tree
149.	Strychnos potatorum Linn. (Loganiaceae)	Clearingnut tree
150.	Symplocos cochinchinensis Roxb. (symplocaceae)	Lodh tree
151.	Syzigium cumini Skeels (Myrtaceae)	Jambolan
152.	Tamarindus indica Linn. (Leguminosae)	Tamarind
153.	Tectona grandis Linn. f. (Lamiaceae)	Teak tree
154.	Terminalia arjuna W. and A. (Combretaceae)	Arjun tree
155.	Terminalia bellirica Roxb. (Combretaceae)	Beleric myrobalan

(Continued)

(*Continued*)

Plant	Botanical name of the plants (family)	Common name
156.	Terminalia chebula Retz. (Combretaceae)	Chebulic myrobalan
157.	Thespesia populnea Sol. (Malvaceae)	Tulip tree
158.	Tinospora cordifolia Miers (Menispermaceae)	Heart-leaved Moonseed
159.	Trachyspermum ammi Sprague (Umbelliferae)	Bishop's weed (Ajowan)
160.	Tribulus terrestris Linn. (Zygophyllaceae)	Small caltrops
161.	Trigonella foenum-Graecum Linn. (Leguminosae)	Fenugreek
162.	Triticum aestivum Linn. (Poaceae)	Wheat
163.	Urginea indica Kunth. (Liliaceae)	Indian squill
164.	Urtica dioica Linn. (Urticaceae)	Stinging nettle
165.	Vateria indica Linn.(Dipterocarpaceae)	Indian copal tree
166.	Valeriana Wallichii (Valerianaceae)	Indian Valerian
167.	Vernonia anthelmintica Willd. (Compositae)	Wild cumin
168.	Viola odorata Linn. (Violaceae)	Common violet
169.	Withania somnifera Dun. (Solanaceae)	Winter cherry
170.	Zingiber officinale Rosc (Zingiberaceae)	Ginger
171.	Ziziphus jujuba Lam (Rhamnaceae)	Indian date

Source and Bibliography

1. Health Secrets and Healing Herbs, Durga Nath Dhar and Pankaj Dhar, Galgotia Publications Private Limited, New Delhi-(2006).
2. Cultivation of Medicinal Plants, C.K. Kokate, A.S. Gokhale and S.B. Gokhale, Nirali Prakashan, Pune (2004).
3. Important Medicinal Plants of India, Arti Tripathi, Utthan, Centre for Sustainable Development and Poverty Alleviation, Allahabad (2003).
4. Natural Remedies, D.N. Dhar and Rupa Dhar, An everyday Guide to Herbal Teas, Infusion and Decoctions, Orient Paperbacks, Delhi/Mumbai/Hyderabad (2002).
5. Herbal Cures for common Ailments, James E.O'Brien, American Media Mini Mags, Inc., Florida, USA, (2000).
6. Doctors Book of Home Remedies for cold and flu, Editors of Prevention, rodale, rodale Inc. (2000).
7. Dental First Aid for Families, R. Diamond, Idyll Arbor Inc., Ravendale, WA, USA (2000).
8. The Arthritis Foundation's Guide in Alternatives, J. Horstman, Arthritis foundation, an official Arthritis foundation (1999).

9. Chinese Herbal Secrets, S. Chmelik, Avery Pub. Group, New York (1999).

10. Common Food Plants and Their Medicinal Properties, Tibetan Healing, Peter Fenton, Quest Books, (1999).

11. A complete Guide to Prevention, Treatment and Healthy Living- Get Healing Now. Gary Null, (1999).

12. Aromatherapy A-Z, Leatham and C. Higley, Hay House, Inc; Carlsbad, CA(1998).

13. The Anxiety Cure, R.L. Dupont, E.D. Spencer and C.M. Dupont, John Wiley and Sons, (1998).

14. Holy Basil, Tulsi (A Herb), Yash Rai, Navneet Publication (1998).

15. Factfile (Simple Techniques for Pain Relief) C.M. McLaughlin, Time Life Books, Alexandria, Virginia (1998).

16. Chamomile and its cultivation in India, Muni Ram, Madan Mohan Gupta, Sushil Kumar, Farm Bulletin No. 002, 8pp Central Institute of Medicinal and Aromatic Plants, Lucknow (1997).

17. The chiropractor's Health Book, L. Mcgill, Crown Trade Paperbacks, New York, (1997).

18. The Handbook of Ayurveda: Kyle Cathic Ltd. (Great Britain), Journey Editions, An Imprint of Charles E. Tuttle co. Inc. (1997).

19. The Four Pillars of Healing. L. Galland, Renaissance Workshop Ltd. (1997).

20. A Russian Herbal I.V. Zevin, Healing Arts Press, Rochester, Vermont, (1997).

21. The Alternative Advisor: The complete Guide of Natural therapies and Alternative Treatments. The Editors of time Life Books, Alexandria, Virginia, (1997).

22. The Encyclopedia of Aromatherapy, C. Wildwood, Healinh Arts Press, Rochester, Vermont, (1996).
23. The knee book, H. Kiernan, Crown Publishers Inc., New York, (1995).
24. Home Remedies What Works: G. Maleskey & B. Kufman, Rodale Press, Emmaus, Pennsylvania, (1995).
25. Natural Health Secrets from Around the World, Editors. G.W. Gealhoed, R.D. Willex, Jr., J. Barilla, Shot Tower Books, Inc. Florida, (1994).
26. The Secret Medicine of the Pharaohs, An Egyptian Healing Cornlius Stretter. Editionsq Chicago, Berlin, Moscow, Tokyo, (1993).
27. Solutions for a Healthy Life, R. Kurzweil and A.M. Gotto, Jr. Crown publishers, Inc., New York, (1993).
28. Herbs that Heal, Natural Remedies for Good Health, H.K. Bakhru, Orient Paperbacks, New Delhi/ Bombay, (1992).
29. Learn Massage in a weekend. N. Lacroix, Alfred A. Knopf, New York, (1992).
30. Eight steps of a Healthy Heart, R.E. Kowablski, Warner Books, (1992).
31. The Healing Benefits of Acupressure, F.M. Houston, Keats Publishing Inc., Connecticut, (1991).
32. Free Yourself from Headaches, J. Stromfeld and A. Weil, New American Library, (1989).
33. How to control Blood Pressure without Drugs, R.L. Rowan, Charles Scrsibner and Sons, New York, (1986).
34. The Whole Body Healing C. Lowe, J.M. Nechas and Editors of Prevention Magazine, Rodale Press, Emmaus, Pa., (1983).

35. The complete Book of Home Remedies, Hakeem H. Abdul Hameed Saheb, Orient Paperbacks, New Delhi, (1982).
36. The Magic of Herbs, David Conway, May flower Books, (1977).
37. The Chinese Art of Healing. S. Palos, Herder and Herder, (1971).
38. Massage at your Finger Tips, A qualified Masseur, Thorsons Publishers Ltd, London (1971).
39. Medicinal Plants of India and Pakistan, J.F. Dastur, D.B. Taraporevala Sons and Co. Private Ltd., Bombay, (1962).
40. Essential Oil Composition of *Pimpinella anisum* L. fruits from various European countries. A. Orav, A Raal, and E.A. Arak. Natural Product Research, 2008, 22, 227–232.
41. A review on phytochemistry and ethnopharmacological aspects of genus Calendula. D. Arora, A. Rani, and A Sharma. Pharmacognosy Reviews, 2013, 7, 179–187.
42. Bioactive potential of *Anethum graveolens*, *Foeniculum vulgare* and *Trachyspermum ammi* belonging to family Umbelliferae. G.J. Kaur and D.S. Arora. Journal of Medicinal Plants Research, 2010, 4, 87–94.
43. Pharmacological Studies of Passiflora sp. and their Bioactive Compounds. A.G. Ingale and A.U Hivrale. African Journal of Plant Science, 2010, 4, 417–426.
44. Phytochemical and Pharmacological Properties of *Gymnema sylvestre*: An Important Medicinal Plant. P. Tiwari, B.N. Mishra and N.S. Sangwan. BioMed

Research International, Volume 2014 (2014), Article ID 830285.

45. Effects of Turmeric on Alzheimer's Disease with Behavioral and Psychological Symptoms of Dementia. N. Hishikawa, Y. Takahashi, Y. Amakusa, Y. Tanno, Y. Tuji, H. Niwa, N. Murakami and U.K. Krishna. Ayu, 2012, 33, 499–504.

Index